Walter Charles Blasdale

A description of some Chinese vegetable food materials and

their nutritive and economic value

Walter Charles Blasdale

A description of some Chinese vegetable food materials and their nutritive and economic value

ISBN/EAN: 9783337201487

Printed in Europe, USA, Canada, Australia, Japan

Cover: Foto ©berggeist007 / pixelio.de

More available books at **www.hansebooks.com**

BULLETIN No. 68.   301

U. S. DEPARTMENT OF AGRICULTURE,
OFFICE OF EXPERIMENT STATIONS.
A. C. TRUE, Director.

A DESCRIPTION

OF

SOME CHINESE VEGETABLE FOOD MATERIALS

AND

THEIR NUTRITIVE AND ECONOMIC VALUE

BY

WALTER C. BLASDALE,
Instructor in Chemistry, University of California.

WASHINGTON:
GOVERNMENT PRINTING OFFICE.
1899.

LETTER OF TRANSMITTAL.

U. S. DEPARTMENT OF AGRICULTURE,
OFFICE OF EXPERIMENT STATIONS,
Washington, D. C., July 15, 1899.

SIR: I have the honor to transmit herewith a report by Walter C. Blasdale, instructor in chemistry at the University of California, describing some Chinese vegetable food materials and their nutritive and economic value.

These foods are used to a considerable extent by the Chinese population in San Francisco and other cities in the United States, and most, if not all, of them are staple articles of diet in China and the Orient. It seems probable that some of the vegetables may become generally and favorably known in the United States.

Very little information has been hitherto available concerning many of these materials, and it is believed the report is a useful contribution to the knowledge of the food of mankind.

The report is respectfully submitted, with the recommendation that it be published as Bulletin No. 68 of this Office.

Respectfully,

A. C. TRUE,
Director.

Hon. JAMES WILSON,
Secretary of Agriculture.

2

CONTENTS.

ILLUSTRATIONS.

4

SOME CHINESE VEGETABLE FOOD MATERIALS.

INTRODUCTION.

A visit to the Chinese quarter of San Francisco or any of the larger Pacific Coast cities will reveal to the eyes of a nonresident much that is both strange and interesting. Most of the curious roots, green vegetables, seeds, and other articles of food making up the stock in trade of the Chinese merchant would be totally unknown and unsalable outside of the narrow limits of the Chinese quarter. These articles are for the most part of Asiatic origin, many of them being directly imported from Canton, while others, though grown on American soil, are distinctly Asiatic in character. Their presence here can be accounted for only on the supposition that they are of considerable importance in the domestic life of the Chinese. Evidently they are the equivalent of the materials that make up our own vegetable dietary and presumably possess an intrinsic value for such a purpose. The thought then naturally arises, Might not some of these materials be turned to good account in the American household?

Tradition assigns to the Chinese the highest attainment in the art of producing from a given area the greatest amount of food material. The latter result has been reached both by intensive cultivation and by the utilization of a great variety of food plants, so that all classes of soil and climate are made to yield their quota of food. It would not be unreasonable to suppose, therefore, that the little-known regions of the Chinese Empire from which we have already obtained many useful plants might yield still others of real economic value. It can not be taken for granted, however, that all such materials, even though they do form an important part of the Chinese dietary, would be desirable introductions into our own. Of the numerous factors which must be considered in determining this question the composition of the vegetables themselves is the most important. The amount of nutrients which they contain may be readily determined by submitting them to a chemical investigation. Such questions as digestibility, adaptation to American tastes, means of utilization, and cultural features of the plants themselves are all important elements in the problem.

Powerful incentives which might induce one to look for new varieties of food plants are not lacking. An increase in the number of vegetables which are in cultivation might admit of the utilization of a greater diversity of soil and climate, or of a more profitable use of the regions

5

already in cultivation, or again might be desirable from a strictly dietetic standpoint.

It has long been accepted as true that the Chinese are largely vegetarian in their diet, and this apparently without serious detriment to their physical development. This statement, however, is not based on accurately compiled data, and, even if true, would be of no special significance until we know more about the composition of the Chinese vegetables. Apparently few analyses of Chinese food materials have been published, though Kellner, Nagai, Murai, and others have published a large number of analyses of Japanese food stuffs, many of which are produced by plants in common use among the Chinese. The incomplete character of some of these analyses and indeed of many others is likely to lead to erroneous conclusions. It is especially desirable to discriminate between crude protein and digestible nitrogenous compounds when dealing with vegetable substances, since often as much as two-thirds of the total nitrogen content consists of amido[1] compounds or other substances of little or no nutritive value.

EXPERIMENTAL METHODS.

It was through the consideration of such facts as the above that the work described in this article was undertaken. It has been confined to a study of the most important of the vegetable food materials found in the Chinese markets of San Francisco, though many substances of animal origin found there might profitably be submitted to similar investigation.

The work naturally divides itself into two distinct lines of research: First, the botanical and horticultural features of the materials studied, and, second, the extent to which these materials are used, the method of preparing and serving them, and the food value as shown by their chemical composition.

As a starting point for both lines of work it was necessary in each case to identify the plant from which the product under investigation was derived.[2] The identification proved in some instances a more difficult matter than might have been anticipated, for, though many notes on the economic plants of China are available, a large number of them contain conflicting statements and several different Chinese characters are often used to designate the same plant. In those instances in which the plants themselves could be grown by the author little difficulty was

[1] In confirmation of this statement see Böhmer, Landw. Vers. Stat., 28 (1883), p. 247.

[2] An especial acknowledgment is due Prof. John B. Fryer, of the department of oriental languages of the University of California, and to Mr. W. N. Fong, a student in the same institution, for assistance in this part of the work and for other courtesies; also to Mr. Charles Ford, director of the botanic garden at Hong Kong, for identification of the roots of *Pachyrhizus angulatus* and for references to works on the botany of China; and to Prof. W. A. Setchell, of the department of botany of the University of California, for the identification of species of algæ.

experienced, but this was in some instances an impossibility. As an assistance in this part of the work and also for the benefit of future workers in the same field, the name and Chinese character for each article were obtained when purchased, and these have been recorded in the body of the report. It is to be noted, however, that the names used in San Francisco are all in the Cantonese dialect, and both names and characters often represent commercial designations rather than the terms used in Chinese literature. In many instances the literary or classical terms have been added. Often the names obtained from the Chinese merchants gave but an uncertain clew to the name of the plant producing the article, and frequently different names were used for the same article. Aside from names and characters it was very difficult, either from their actual ignorance or lack of interest in the subject, to obtain authentic information from the Chinese consulted by the authorities in San Francisco regarding the source and method of use of many of the substances examined.

On the chemical side of the work the chief difficulty encountered was in the choice of the method of analysis to be employed. The uncertainties involved in many of our present methods of food analysis are too well known to need any extended mention here. In order to make the results comparable with other investigations of a similar nature, the methods outlined by the Association of Official Agricultural Chemists [1] have, for the most part, been adhered to and only such minor modifications have been introduced as the nature of the work seemed to demand.

The substances as purchased in the markets were weighed as soon as possible, and, in the case of fleshy vegetables, the water content reduced by drying in an air bath whose temperature did not exceed 70° C. The residue was allowed to absorb what moisture it would, was then weighed again, and the percentage of loss calculated. This partially dried material was then ground either in a coffee mill or by means of an agate mortar until it all passed through a sieve with 50 meshes to the linear inch and was immediately placed in a tightly stoppered bottle and used for the various determinations. The residual water was carefully determined both at the beginning and at the end of the analysis by drying in a water bath until the sample ceased to lose weight. In many instances the time occupied in making the entire analysis extended over a period of eighteen months, and some samples showed differences of as much as 2 per cent in the two determinations. In such cases the average of the two determinations was obtained and this average factor was used in reducing the results of the analysis to the figures corresponding to a water-free basis. A few exceptions to this method of procedure are noted in the account of the results of the individual analyses.

[1] U. S. Dept. Agr., Division of Chemistry Bul. 46.

For the determination of the crude protein the Kjeldahl method as adopted by the Association of Official Agricultural Chemists was employed. Albuminoids were determined by the Stutzer method, and, with the exception of the difficulty encountered in filtering and washing the cupric hydrate precipitate, this proved fairly satisfactory. This difficulty was particularly marked in those samples which contained large amounts of starch. In such instances great care had to be exercised to avoid overheating, as this gave a precipitate of such gelatinous character that filtration became impossible.

Various attempts to determine the amido compounds directly, by digestion with 5 per cent hydrochloric acid and distillation of the resulting solution with barium carbonate or magnesium hydrate, either failed to give the entire amount of the amids or were unsatisfactory in the presence of albuminoids. The method of Sachsse was found equally unsatisfactory. Hence it was necessary to report simply the difference between the figures found for crude protein and those found for albuminoids, as "amids by difference."

Although the provisional methods recommended by the association for determining the various members of the carbohydrate group are not entirely satisfactory, they have been followed in general. The results are believed to be more useful than a determination of the whole group by difference.

The determination of crude fiber has been made according to the official method, except that asbestos was used in both filtrations. The method gave concordant results when applied to the same sample, but, owing to the varying degrees of difficulty experienced in making the filtrations, can not be applied to a series of different materials under exactly similar conditions. In applying this method to the two species of fungi analyzed nearly two days were required for both filtrations, during which time the acid and alkali solutions were in contact with the material and must have caused lower results than would have been obtained had the method been carried out under normal conditions.

A total of 53 samples was collected, 42 of which were analyzed. For the sake of convenience they have been divided into a number of groups in reporting and discussing the work.

ROOTS AND TUBERS.

SAGITTARIA, OR ARROWHEAD.

The use of the autumn tubers of the various species of Sagittaria, or arrowhead, as articles of food is a common practice among several different races. Lewis and Clarke, in the account of their memorable journey of 1804 across the western United States, make frequent mention of the use of the tubers of *Sagittaria latifolia* (called Wappato) by the Chinook Indians of Oregon. Kalm[1] refers to the use of the roots

---

[1] Pinkerton's Voyages, 13, p. 523.

PLATE I.

PLANT FROM AN IMPORTED TUBER OF *Sagittaria sinensis*.

of a species of Sagittaria by the Indians of eastern America. Parry[1] notes the use of the tubers of *S. latifolia* by the Chippewa Indians, and states that in a raw state they contain a bitter, milky juice, but on boiling become sweet and palatable. Finally Coville[2] notes the use of the tubers of *S. arifolia* and *S. latifolia* by the Klamath Indians of Oregon. References to the use of *S. sagittifolia* and *S. sinensis* throughout China, Japan, and portions of India are not uncommon in books of travel relating to these countries, and apparently forms of these species have been in cultivation in the East from very early times. Henry[3] states that the arrowhead (*S. sagittifolia*) is cultivated in all parts of China for its edible tubers, and that there is also a wild form with numerous varieties which produce smaller tubers. The wild form occurs in almost every pool and paddy field. Smith[4] refers to the use of Sagittaria, but the inference from his statement is that it is not very widely cultivated.

The tubers of two species of Sagittaria are found in the Chinese market of San Francisco. Those of *S. latifolia* are not uncommon from October to May, and are known to the Chinese as "t'sz ku" or "chu ko," the former name being the classical one and the latter a provincialism. The characters for them are 茨 菇, but they differ from the characters used in Chinese literature. These tubers are the product of the native American plants which grow abundantly on the "tule lands" bordering the Sacramento and San Joaquin rivers, but, though more largely used, apparently are not esteemed as highly as those of the Chinese species mentioned below. The tubers average about 6 centimeters in length by 2 to 4 in diameter, and weigh about 14 grams each. They are often somewhat flattened laterally and are terminated by an elongated sprout which generally exceeds the length of the tuber itself. In color they are tawny white with often a pronounced bluish tinge, the color being due solely to the very thin, smooth skin. This is crossed by from two to four encircling lines which mark the position of axillary buds. The interior is of about the same consistency as a potato, yellowish in color, and farinaceous in taste.

Well-developed plants were successfully grown from both forms of tubers, which were obtained at different seasons and from different vendors, though none of these plants yielded mature fruit. The method of formation of the tubers is somewhat peculiar. The elongated terminal sprout rapidly develops and soon produces a terminal bud, from which at first roots and then leaves are produced, thus forming an independent plant. This in turn sends out a new series of runners, which behave in the same manner. The original tuber either decays at once or may

---

[1] Plants of Wisconsin and Minnesota, Owen's Report, 1852, p. 619.

[2] Contrib. U. S. Nat. Herb., 5 (1897), No. 2, p. 90.

[3] Notes on the Economic Botany of China, p. 27. Shanghai, 1893.

[4] Contributions toward the Materia Medica and Natural History of China, p. 189. Shanghai, 1871.

10

form more shoots from some of its numerous axillary buds, which in turn give rise to other plants. Toward autumn all the plants which have been thus established develop long runners, each of which bears a single terminal tuber and thus completes the life cycle without the introduction of any sexual process. These features render the plant remarkably prolific.

In the present investigation the author has been unable to study our native Sagittaria with any degree of thoroughness, but it is certain that there are constant differences between the native and imported tubers and the respective plants which may be grown from them. Plate I shows a plant grown from one of the imported tubers. The differences in the two forms of tubers are shown in Pl. II, figs. 3 and 5.

The specific limitations of the members of this genus are not well defined, but there can be but little doubt that the first form of tuber studied is the product of the different varieties of *S. latifolia*, while the second form, in so far as can be determined in the absence of mature akenes, seems to be the product of *S. sinensis*. Both of these species, however, are sometimes regarded as synonymous with *S. sagittifolia*. The leaf characters of the plants produced by the first variety vary considerably, while those produced by the second form are fairly constant. The former commonly produce leaves 1 meter in length with smooth, slender, semiterete petioles, having obtusely pointed apices and divergent lanceolate lobes; also flowers that agree in all essential respects with those of *S. latifolia*. The plant is beautiful, and is well worthy of a place among the ornamental aquatics. Indeed, it is frequently used for this purpose by the Chinese, who grow it in vessels of water in the same manner as they do the Chinese narcissus.

The plants produced by the second form of tuber studied have slightly smaller leaves, with distinctly five-ribbed petioles, a very obtuse apex, and more pronouncedly divergent lobes. The plant has a more stocky habit than the former species, and the tuber-bearing runners are much shorter.

The statement is made by Parish[1] that *S. sinensis* has been introduced by the Chinese into southern California, and by Brandegee[2] that it is common along the Sacramento River. Both these statements, however, are disputed by Smith,[3] and apparently on sufficient grounds.

The tubers of *S. sinensis* appear in the markets in December or January, in some seasons in large quantities, and as far as could be learned are always imported from Canton. They are symmetrical and nearly spherical in form, but in size, color, consistency, and other characteristics closely resemble those of the American species and the same Chinese character and name are used for them.

[1] Zoe, 1 (1890), p. 122.
[2] Zoe, 4 (1893), p. 247.
[3] Revision of North American Species of Sagittaria and Lophotocarpus, 1894, pp. 9, 12.

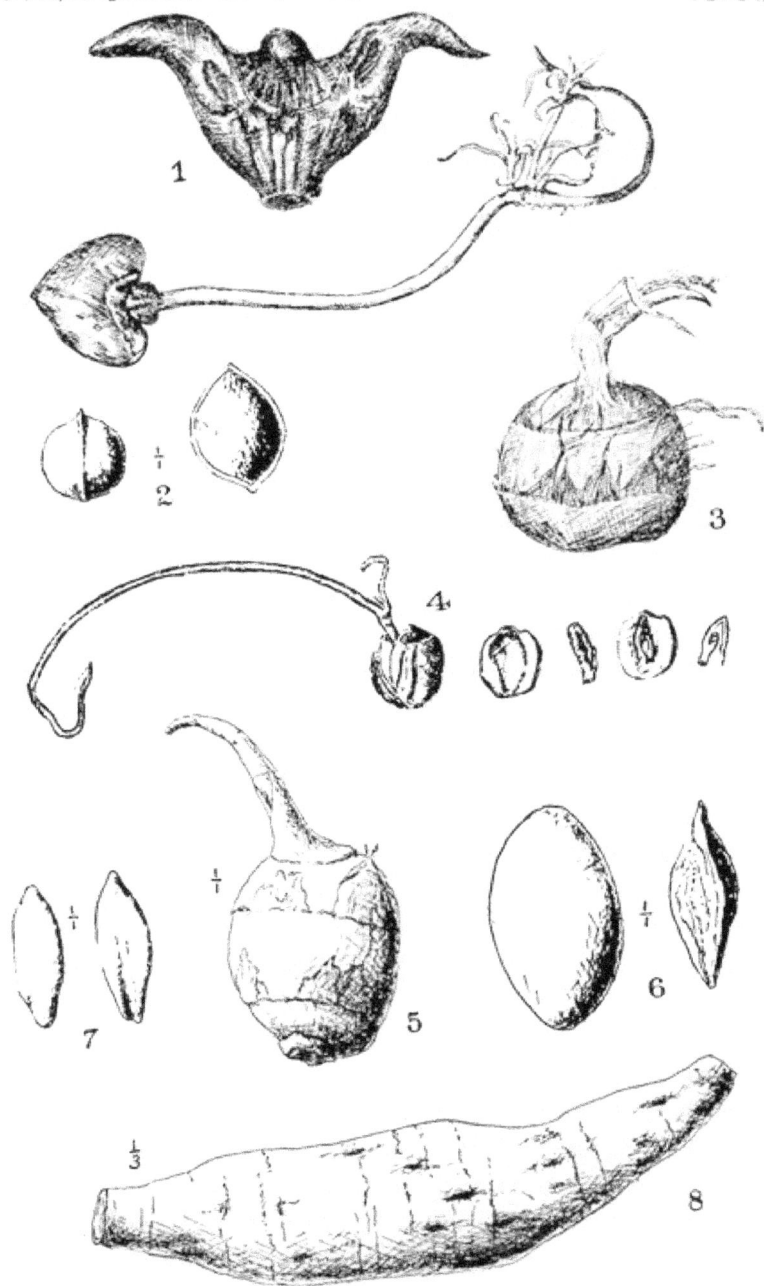

DRAWINGS OF ROOTS, SEEDS, ETC.

1. Seed and young plant of the "horn chestnut" (*Trapa bispinosa*); 2. Seed of *Ginkgo biloba*;
3. Tuber of *Sagittaria sinensis*; 4. Seed and young plant of *Nelumbium speciosum*; 5. Tuber
of *Sagittaria latifolia*, sprouted; 6. Fruit and seed of the Chinese olive (*Canarium album*);
7. Kernel of Canarium seed; 8. Root of *Pueraria thunbergiana*.

The starch grains of both species vary from orbicular to ovate in their plane of greatest extent, are occasionally somewhat angled, and seldom exceed 30 μ in diameter (Pl. III, fig. 1). The hilum is somewhat eccentric and the striations are pronounced. With polarized light and a selenite plate they exhibit a slight play of colors.

The composition of the two sorts of tubers is shown in Table 1. The only previous analysis of this vegetable which has been found is reported by Kellner, and represents a Japanese sample of the tubers produced by *S. sagittifolia*. This is included in the table for purposes of comparison.

TABLE 1.— *Composition of arrowhead tubers.*

	Water.	Protein.	Albuminoids.	Amids (by difference).	Fat.	Starch.	Cane-sugar.	Pentosans.	Crude-fiber.	Ash.	Undetermined.
	P. ct.	P. ct.	P. ct.	P. ct.	P. ct.	P. ct.	P. ct.	P. ct.	P. ct.	P. ct.	P. ct.
Sagittaria latifolia:											
Original material....	66.88	4.44	3.98	0.46	0.76	19.69	a 2.49	......	0.98	2.04	2.71
Water-free substance	13.41	12.02	1.39	2.29	59.46	7.51	......	2.97	6.17	8.19
Sagittaria riensis:											
Original material....	61.51	7.00	4.71	2.29	.24	22.95	a 2.26	0.32	.72	1.69	3.31
Water-free substance	18.18	12.24	5.94	.62	63.89	5.87	.93	1.87	4.38	7.26
Sagittaria sagittifolia: b											
Original material ...	66.86	7.05	5.76	1.29	.55	c 22.93	......	......	4.14	1.45	......
Water-free substance	21.26	17.38	3.88	1.66	c 69.21	......	......	3.56	4.31	......

a Sample contained no reducing sugars.
b Reported by Kellner, quoted from König Chemie der menschlichen Nahrungs- und Genussmittel, 3. ed., 1, p. 705.
c Carbohydrates by difference.

A comparison of the different analyses shows that the figures obtained by Kellner differ little from the results obtained in this laboratory, and in no instance are the differences greater than might have been looked for in tubers grown in different regions. All three analyses show that these tubers contain considerably less water than the majority of our commonly cultivated root crops, and, like most such foods, only small amounts of fat, ash, and crude fiber.

The most striking feature is the high content of crude protein, which, generally speaking, is considerably higher than that of potatoes and similar vegetables. Kellner's analysis gives only total nitrogenous substance, but the other two show a remarkably high percentage of albuminoids, the difference in the crude protein of the two analyses being due almost entirely to nonalbuminoid nitrogenous matter; so that, judging by these analyses, the two forms are about equally valuable as far as albuminoid nitrogen is concerned. The superiority of the arrowhead over the potato in this respect is still more striking when it is remembered that about 50 per cent of the nitrogenous content of the latter vegetable is of a nonalbuminoid character.

Of the carbohydrate constituents, starch constitutes by far the most important part, though a considerable amount of some member of the cane-sugar group is present. In one analysis a small amount of some reducing sugar was found, but this may have been due to a slight inversion of some body of the cane-sugar group. Furfurol-yielding compounds are also present, though only in small amounts.

Some attempt was made to determine the exact nature of the nitrogenous compounds present. An aqueous extract of the dry residue of the second sample in the table yielded 15.66 per cent of nitrogenous substance; hydrochloric acid added to this extract precipitated 5.96 per cent of the original substance, 97 per cent being of a nitrogenous character. An extract of the freshly crushed tubers gave a precipitate with both potassium ferrocyanid and with phosphomolybdic acid. On heating, this extract gave a heavy coagulum at 74° C., but none above that temperature, though the filtrate from the coagulum still gave a precipitate with both the potassium ferrocyanid and the phosphomolybdic acid. With hydrochloric acid the extract gave a heavy precipitate soluble in a large excess of the reagent. With cupric hydrate it gave a blue solution and a heavy precipitate. When the filtrate from the extract which had been coagulated by heat was evaporated somewhat and allowed to stand, several large colorless crystals resembling in form those of asparagin were obtained. These also gave the well-known reactions of that compound. Hence, it would seem that nitrogenous constituents of the arrowhead tubers consist of (1) a soluble casein-like albuminoid comprising nearly 6 per cent of the dry tubers, (2) a soluble albumen coagulating at 74° C., (3) a small amount of an insoluble albuminoid of an unknown character, and (4) varying amounts of nonalbuminoids of which asparagin is an important constituent.

The total aqueous extract of the dry tubers (the second sample in the table) amounted to 29.18 per cent of the entire weight. Subtracting from this figure those corresponding to the soluble nitrogenous constituents, the soluble ash and the soluble carbohydrates, there remains 5.28 per cent of a soluble substance unaccounted for. The aqueous extract has a peculiar but not unpleasant odor. After coagulating the albuminoids, filtering and evaporating to dryness, there remained a brown sticky mass, very sweet in taste and having in a pronounced degree the peculiar odor of the extract itself. Hence, there is probably at least one unidentified substance which gives to the tubers their peculiar taste. This could not be obtained in a pure condition.

The statement sometimes made that the arrowhead has acrid properties apparently refers only to the leafy portion of the plant, or has arisen from a confusion with some other cultivated water plant.

The foregoing facts seem to warrant the opinion that the arrowhead offers many advantages from a dietetic standpoint. When cooked in the same manner as potatoes, it forms, in the author's opinion, a very acceptable food, though it is not without a pronounced and characteristic flavor. It is stated that some of the poorer non-Mongolian residents of San Francisco have acquired a liking for it and use it freely. The utilization of the arrowhead has already been commented on favorably by

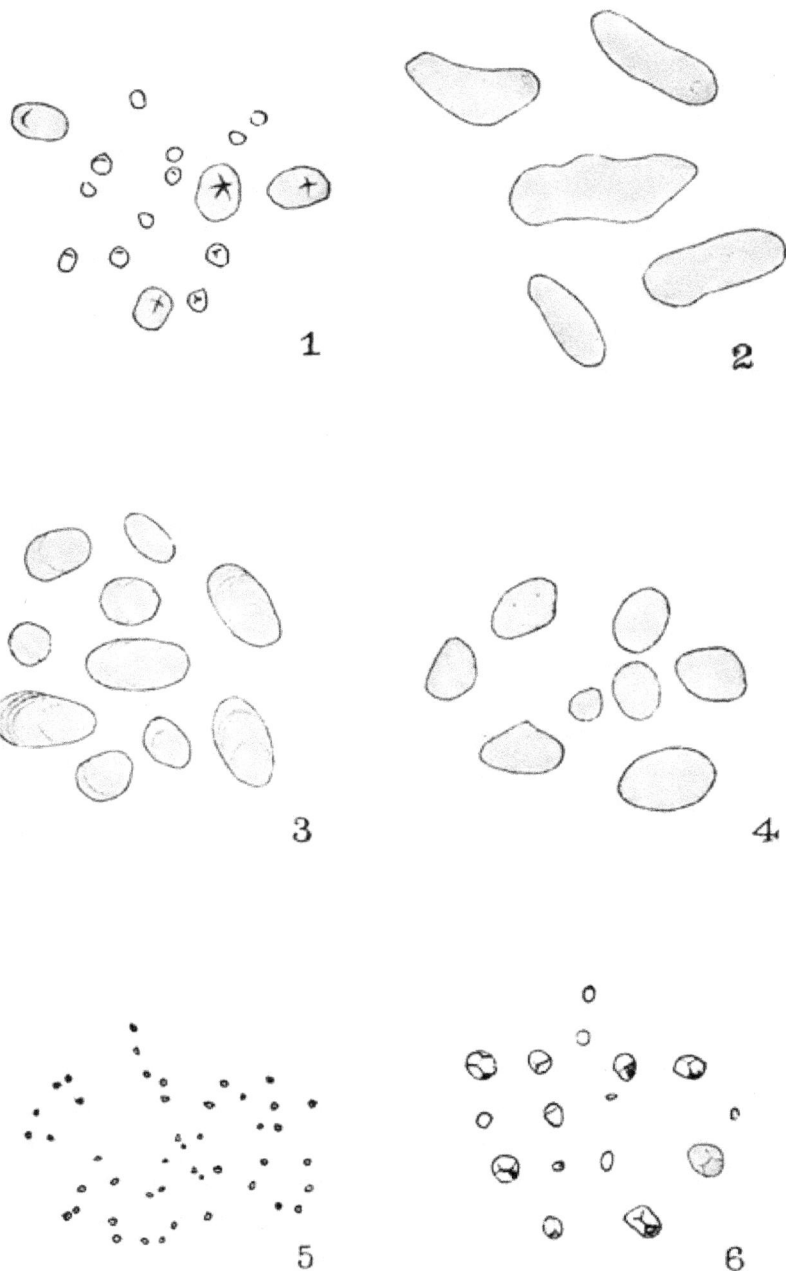

1

2

3

4

5

6

DRAWINGS OF STARCH GRAINS MAGNIFIED ABOUT 480 DIAMETERS.

1, *Sagittaria sinensis*; 2, *Nelumbium speciosum* (roots); 3, *Lilium japonicum brownii*; 4, *Trapa bispinosa*; 5, *Euryale ferox*; 6, *Amorphophallus rivieri*.

·

Pailleaux and Bois.[1] The tubers might also be used in the manufacture of starch. They are employed for this purpose by the Chinese, and it is said a fine quality of that article is easily prepared from them by the usual methods. The comparatively large areas of land both in California and in other parts of the United States which are too wet for the cultivation of other root crops would seem to offer an especial inducement to considering the utilization of this genius of plants.

TARO.

From remote antiquity the taro has furnished an important item of the food supply of the natives of southern India, Australia, portions of Africa, and many of the islands of the Pacific, and is to-day one of the plants most commonly cultivated throughout the Tropics. Botanically speaking, the taro cultivated for food may be any one of a number of species of the genus Colocasia (also known as Caladium), representatives of the family Aroidæ. The species commonly designated as *Colocasia antiquorum*, of which a large number of varieties are known, seems to be the one most widely cultivated, though *C. indica*, *C. odora*, and *C. macrorhiza*, the two former possibly only varieties of *C. antiquorum*, are said to be extensively used. Most of these species are also used as ornamental greenhouse plants in temperate climates.

Two forms of the taro are sold in large quantities in the Chinese market of San Francisco. Both are said to be imported either from Canton or from the Sandwich Islands. The first, which is the smaller form, is designated by the Chinese characters 芋仔, the second by 扮薯芋. The two forms differ only in size, and produce plants which do not show even varietal differences. Evidently they are the roots of *C. antiquorum*. The smaller form consists of small ovoid or ellipsoid roots about 10 centimeters in length, which weigh from 60 to 70 grams. The larger roots are about 24 centimeters in length, and weigh from 500 to 800 grams. The latter form is considered superior to the former by the Chinese and is sold at a slightly higher price.

The interior of the roots has about the consistency of a sweet potato, and a microscopical examination reveals the presence of large amounts of starch, which is present in the form of exceedingly small grains.

The roots are easily started into growth, and the Chinese market furnishes a cheap source of supply of this well-known ornamental plant.

Several analyses of species of Colocasia have been made by different investigators, most of which, however, are somewhat incomplete. In Table 2 are shown analyses of the two forms studied in this laboratory, one of a small and one of a large root. For purposes of comparison, an analysis by Kellner and one by Nagai and Murai are quoted:

[1] Soc. d'Acclimat., 4. ser., 5 (1888), p. 1102.

TABLE 2.—*Composition of taro.*

	Water.	Protein.	Albuminoids.	Amids (by difference).	Fat.	Starch.	Cane sugar.	Reducing sugars.	Crude fiber.	Ash.	Undetermined.
Colocasia antiquorum (small root):	*Per ct.*	*Per ct.*	*Per ct.*	*Per ct.*	*Per ct.*	*Per ct.*	*Per ct.*	*P. ct.*	*P. ct.*	*P. ct.*	*P. ct.*
Original material ...	74.20	1.79	1.67	0.04	0.27	17.95	1.15	.00	0.98	1.31	2.44
Water-free substance..............	......	6.69	6.46	.14	1.04	69.58	4.45	.00	3.78	5.10	9.45
Colocasia antiquorum (large root):											
Original material ..	67.51	1.89	1.62	.27	.16	23.32	1.86	.00	.66	1.10	3.30
Water-free substance.............	......	5.51	4.72	.79	.48	73.82	5.43	.00	1.92	3.21	9.63
Colocasia antiquorum; a											
Original material ...	80.65	2.09	1.39	.70	.17	6.52	......	......	.70	.85	15.54
Water-free substance..............	......	10.81	7.18	3.63	.91	33.70	......	......	3.63	4.41	80.24
Colocasia antiquorum; b											
Original material ...	85.20	1.43	......	......	.08	10.40		0.12	......	.99	1.78
Water-free substance..............	......	9.73	......	......	.54	70.26		.81	......	6.69	12.02

a Reported by Kellner, Landw. Vers. Stat., 30 (1884), p. 42.
b Reported by Nagai and Murai, quoted from König, Chemie der menschlichen Nahrungs- und Genussmittel, 3. ed., I, p. 704.

It is difficult to account for the remarkable difference between the starch content reported by Kellner and that reported in the three other analyses. Both the analyses reported from this laboratory were made in duplicate, with substantially the same results. Kellner states that the roots contain large amounts of reducing sugars. The extremely minute size of the starch grains present renders the separation of soluble from insoluble carbohydrates a very difficult matter. The water extract always showed a slight turbidity, even after several filtrations, and it is possible that a considerable part of the substance reported as cane sugar was really starch in a very finely divided condition.

The protein content shows a considerable variation in the four samples. In all cases it is rather low, but even in the two samples from San Francisco, which give the lowest results for this constituent, the amount present is not far from the average of the results obtained for potatoes and similar vegetables. The analyses also show that a large percentage of the crude protein is of an albuminoid nature, the proportion being somewhat greater than in the case of the potato. Most of this was found to consist of a soluble albumin which coagulates at 74° C. A red coloring matter is present in the taro in considerable amounts.

As a whole the taro is characterized by the high percentage of carbohydrates, of which starch is the most important representative, and by the low percentage of fat and crude fiber. From a dietetic standpoint it apparently offers no especial advantage over our commonly cultivated vegetables, but where it is eaten it seems to be a satisfactory substitute for them. It is favorably known to the Anglo-Saxon residents of the Tropics, who soon acquire a liking for it, and even in San

U. S. Dept. of Agr., Bul. 68, Office of Expt. Stations.

PLATE IV.

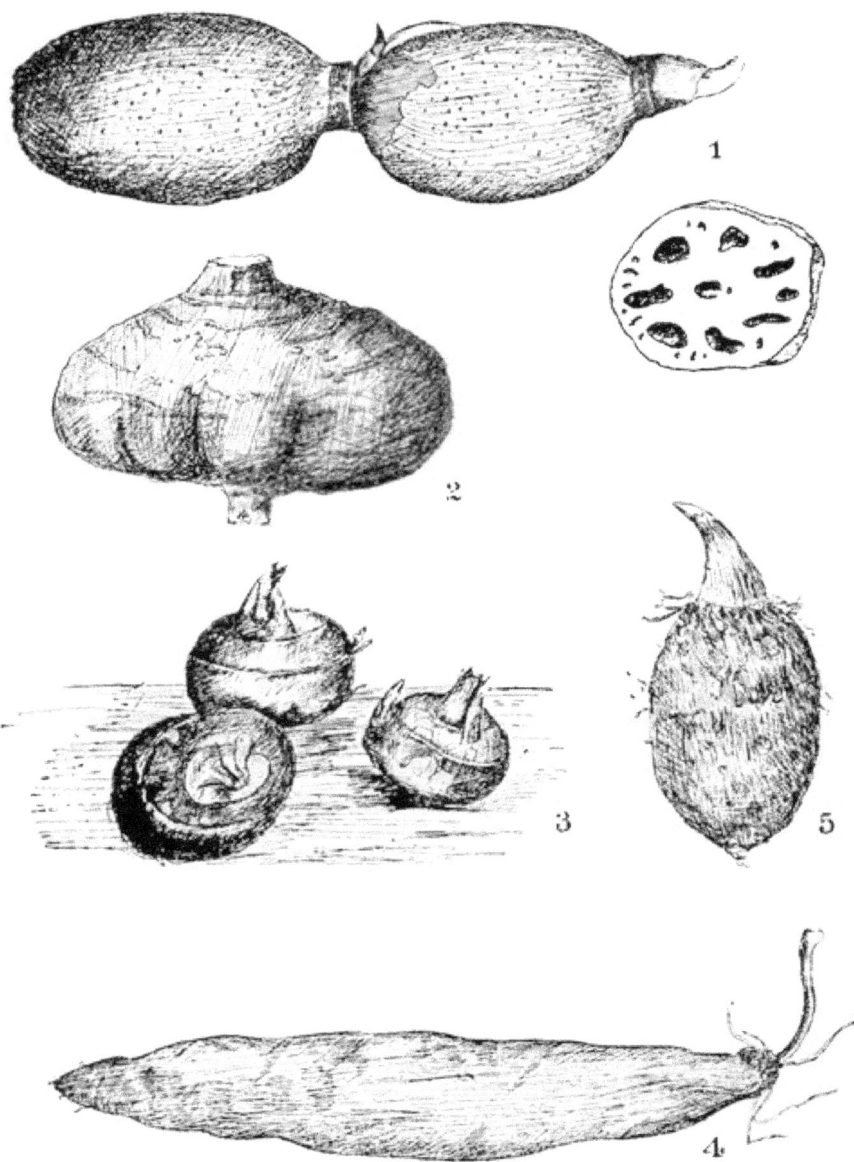

DRAWINGS OF VARIOUS ROOTS.

1. Portion of the rootstock of *Nelumbium speciosum* and cross section of same; 2. Tuber of *Pachyrhizus angulatus*; 3. The water chestnut (*Eleocharis tuberosa*); 4. A root of Manihot, probably *M. utilissima*; 5. A small form of taro (*Colocasia antiquorum*).

Francisco there is a limited demand among the white races for the roots. A root of *C. antiquorum* is shown in Pl. IV, fig. 5.

The area in the United States suited to the culture of taro is somewhat limited. A patch of it has been growing in the Garden of Economic Plants of the University of California for a number of years, and has produced an abundance of large sound roots. In southern California it makes a rapid growth, but requires an abundant supply of water. The Florida Experiment Station[1] has also experimented with it and reports satisfactory results.

WATER CHESTNUT.

Though several species of Scirpus and the allied genus Eleocharis bear tuber-like roots, they are not commonly considered plants of economic importance by Europeans, and are seldom mentioned outside of the systematic manuals. The water chestnut (*E. tuberosa*) is said to be widely cultivated in China and Japan as a food plant, and is there highly esteemed. In the former countries it is known to the English-speaking residents as the water chestnut, which is but a free translation of "ti leh," one of its Chinese names. Smith[2] says that the tubers are largely cultivated and sold for food all over China. "They grow wild in Hupeh in watery places, and are not often especially planted. They are sweet, juicy, and resemble the chestnut in flavor. * * * A kind of arrowroot [i. e., starch] is made from them." It has also been stated[3] that *Scirpus tuberosus* is cultivated in paddy lands for the sake of its tubers, which are eaten either raw or boiled. Bretschneider[4] states that it is cultivated all over China.

The corms of the water chestnut, for they can scarcely be called tubers, begin to arrive in San Francisco markets in the latter part of May, and are to be found on sale throughout the remainder of the year. They are slightly flattened, about 5 centimeters in diameter, and average about 13 grams in weight. They are surrounded by a rather thick chestnut-brown skin, but within are nearly white and of a somewhat watery consistence. Under favorable conditions those purchased in San Francisco will grow readily and produce an abundance of slender terete culms terminated in some instances by many-flowered spikes. Plate V shows one of these plants. In taste the corms are exceedingly sweet, and also possess a peculiar but not unpleasant flavor. Their only resemblance to the chestnut is in the color of the outer skin. To the Chinese they are known as "ma hai," and are designated by the characters 馬蹄, which differ from any of those used by Bretschneider for

---

[1] Florida Sta. Rpt. 1896, p. 9.

[2] Contributions Toward the Materia Medica and Natural History of China, p. 92. Shanghai, 1871.

[3] Useful Plants of Japan, p. 27. Tokyo, 1895.

[4] Jour. China Branch Roy. Asiatic Soc., 25 (1890-91), p. 47.

the plant. A drawing of three water chestnuts is shown in Pl. IV, fig. 3.

As far as observed no analysis of this vegetable has been published. The composition of two samples purchased at different seasons is recorded in Table 3.

TABLE 3.—*Composition of water chestnut.*

	Water.	Proteins.	Albuminoids.	Amids (by difference).	Fat.	Starch.	Cane sugar.	Reducing sugars.	Crude fiber.	Ash.	Undetermined.
Eleocharis tuberosa (first sample):	Per ct.	Per ct.	Per ct.	Per ct.	P. ct.	Per ct.	Per ct.	Per ct.	P. ct.	P. ct.	P. ct.
Original material...	77.29	1.53	1.16	0.37	0.15	7.34	6.35	1.94	0.94	1.19	3.28
Water-free substance		6.73	5.10	1.63	.64	32.30	27.94	8.56	4.12	5.24	14.47
Eleocharis tuberosa (second sample):											
Original material...	77.89	1.31	1.00	.31	.27	8.09	6.02	2.60	1.22	1.18	1.42
Water-free substance		5.91	4.54	1.37	1.23	36.58	27.23	11.78	5.53	5.32	6.42

Much difficulty was experienced in preparing the corms for analysis owing to the high percentage of soluble carbohydrates. This caused them to form a sticky mass, rendering it impossible to reduce the sample to a finely divided condition. Consequently the figures reported for soluble carbohydrates in the analysis of the first sample are probably somewhat low, as is indicated by comparison with the analysis of the second sample in which this difficulty was largely overcome.

On igniting the ash the green color, characteristic of manganese, was always observed. This color was also noted in a large number of the vegetable products grown in China.[1]

The protein content of the water chestnut is rather low, being even lower than that obtained for the taro, but in respect to this constituent, it will still compare favorably with many of the vegetables in common use. Most of the protein consists of soluble albumin which coagulates at $70°$ C. The striking feature of both analyses reported is the high percentage of soluble carbohydrates. Presumably these consist very largely of cane sugar, though several attempts to prepare this in crystalline form from water extracts of bulbs were unsuccessful.

As an article of food the roots, in the author's opinion, are very palatable. The high percentage of sugar gives them an agreeable flavor even while raw, in which state they are commonly used by the Chinese.

But little information is accessible regarding the climatic conditions favorable to the development of the plant. The few specimens which were grown under glass seemed somewhat tender, but it is probable

[1] This would seem to indicate that too much reliance should not be placed on this test as a means of detecting adulterations in teas.

PLATE V.

PLANT FROM A CORM OF *Eleocharis tuberosa*.

that the moister regions of the Southern States or of the interior valleys of California would prove favorable to it. Experiments on this point seem desirable.

SACRED LOTUS.

This plant is better known from the pages of the early classic authors and from tradition than through modern works on economic botany. A plant of such surpassing beauty, whose habit of growth is so unique, could not fail to excite the admiration of all who were permitted to watch its development; and to the people of Egypt and many of the Asiatic countries it assumed the importance of a sacred emblem which was intimately associated with their religion and poetry. In addition to these more æsthetic uses, the plant has from time immemorial been one of economic importance and to-day occupies a not insignificant place in the domestic life of several nations.

One of the most recent accounts of the plant, as viewed in its economic relations, is by Jules Grisard,[1] from which the following facts have been gleaned.

The lotus (Nelumbium) is a native of western India, Persia, Cochin China, and perhaps of Australia. growing in ponds, stagnant water, and rivers. It is cultivated in the basin of the Mediterranean, where there is a summer temperature of about 21° C. In Cochin China it is cultivated in great vases placed at the doors of the houses.

The uses of the lotus are numerous and varied. The stamens are employed in China as an astringent remedy and also for the toilet; the petioles and peduncles furnish a viscous sap employed in India as a remedy for vomiting and diarrhea; the fibro-vascular bundles of the petioles are made into lamp wicks and the carpophore furnishes a popular remedy for blood spitting. The seeds contain a white starch used largely as food (see p. 39); roasted and ground they served the Egyptians for the manufacture of a kind of bread; in China they are used in soup and also as a remedy for indigestion. They are supposed to have invigorative properties when used as food by convalescents. The Chinese also extract from the root a starch which they say is very strengthening. A decoction of the rhizomes is used as a remedy for intestinal inflammation and the rhizomes themselves also become an important article of food in times of famine.

Bretschneider[2] gives an interesting account of the references to the various parts of the plant in the Chinese classics, and Smith[3] gives many notes regarding its varied uses.

The roots or, more strictly, rhizomes of the lotus are brought to San Francisco from Canton in considerable quantities and are on sale in

[1] Soc. d'Acclimat., 1896, p. 189.
[2] Jour. China Branch Roy. Asiatic Soc., 25 (1890), p. 216.
[3] Contributions Toward the Materia Medica and Natural History of China, p. 139. Shanghai, 1871.

the Chinese quarter throughout the late winter and early spring months. As there found, the yellow nodular root stocks are often 100 centimeters in length, each internode measuring from 10 to 20 centimeters in length by about 7 in diameter. The interior is fleshy, but firm, and of a reddish color. The roots are traversed from end to end by a series of ten or more large radially-arranged tubes, with a number of smaller ones between them. The appearance of one of these roots and of a cross section is shown in Pl. IV, fig. 1. The roots contain an abundance of starch grains, which are oval, elliptical, or narrowly oblong, and exhibit a beautiful play of colors with polarized light and a selenite plate. The stratifications are pronounced.

In San Francisco the roots are known as "lin ngau," though the classical name for the plant is "lien hua." The Chinese characters for them are 蓮藕. As far as could be learned, the roots are used in San Francisco for the preparation of a kind of "arrowroot" (i. e., starch), though several authors state that in China the roots are also boiled and used as a vegetable, or are eaten raw in somewhat the same manner as we use salad plants.

It was found impossible to raise plants from any of these roots purchased in San Francisco. The terminal buds in most instances had been destroyed and the roots seemed to be lacking in vitality. There can be no doubt, however, that the roots in question were those of *Nelumbium speciosum.*

In Table 4 are tabulated two analyses of these roots—one of a sample purchased in San Francisco, and the second, which is included for purposes of comparison, of Japanese roots *N. nucifera* [*speciosum*] reported by Kellner.

TABLE 4.—*Composition of lotus roots.*

	Water.	Protein.	Albuminoids.	Amids (by difference).	Fat.	Starch.	Cane sugar.	Reducing sugars.	Crude fiber.	Ash.	Undetermined.
Nelumbium speciosum:	Per ct.	Per ct.	Per ct.	Per ct.	P. ct.	Per ct.	Per ct.	Per ct.	P. ct.	P. ct.	P. ct.
Original material	84.26	1.57	0.94	9.66	0.19	7.71	0.33	2.18	0.76	0.76	2.24
Water-free substance		9.95	5.79	4.17	1.18	48.59	2.12	13.88	4.83	4.81	14.28
Nelumbo nucifera:a											
Original material	85.84	1.09	.73	.36	.20	b 11.14			1.02	.71	
Water-free substance		7.75	5.19	2.56	1.41	b 78.59			7.20	5.01	

a Reported by Kellner, quoted from König, Chemie der menschlichen Nahrungs- und Genussmittel, 3. ed., I, p. 705.
b Carbohydrates by difference.

The two analyses are remarkably concordant. The water content is considerably higher than that of the vegetables previously mentioned. The protein content is rather low, and of this but little more than half is in the form of albuminoids. The latter fact is not surprising when it is remembered that Kinoshita[1] found about 2 per cent of asparagin in the dry roots of *N. nucifera*. The most valuable ingredient of the root, however, is the starch, which constitutes nearly 50 per cent of the dry substance.

A mucilaginous gum manifests itself whenever the fresh roots are cut, and probably this constituent gave rise to the large amount of undetermined matter found. A red coloring matter, whose nature is unknown, was also present in such amounts as to color deeply the water in which pieces of the root had been boiled.

Though the chemical analysis shows that the rhizomes contain valuable food ingredients, their other qualities are such as are likely to prevent them from forming a satisfactory food according to American standards. The roots are decidedly tough and fibrous and somewhat insipid in taste. Long-continued boiling failed either to soften them or bring them into such a condition that they would be easily digested.

The plant might in all probability be utilized for the production of starch, but whether the product would be equal or inferior to the maranta or tacca starch is not known. The form of the starch grains of *N. speciosum* is shown in Pl. III, fig. 2.

The medicinal properties which have been attributed to the various parts of the plant may be seriously questioned, as the Chinese materia medica presents too many incongruities to permit of placing any great reliance on its teachings.

Experiments in the cultivation of the lotus as an ornamental aquatic have been in progress in various parts of the United States for many years and have been unexpectedly successful. It is found to tolerate the severe winters of the Middle East, and the short but hot summers give it an ample season in which to at least perfect its flowers. It is stated that the plant has become naturalized in a small pond near Bordentown, N. J.[2] California and the Southern States afford many regions that are especially suited to the growth of the plant. In the great interior waterway of the former, comprising the Sacramento and San Joaquin rivers, it could scarcely fail to be successful. Strange to say, this species has proved rather more robust than our own native *N. luteum*, whose seeds and roots, it is said, were utilized as articles of food by the aboriginal races of North America.[3]

LILY BULBS.

Though we are accustomed to consider lilies as plants of ornamental value only, the bulbs and flowers of several species have long been used as articles of food. Nitobe[4] gives a very interesting account of the

[1] Imp. Univ. Col. Agr. [Tokyo] Bul., Vol. 2 (1895), p. 203.

[2] The Culture of Water Lilies and Aquatics, Peter Henderson & Co., p. 29.

[3] Engelmann, Trans. St. Louis Acad. Sci., 2 (1860); Newberry, Food and Fiber Plants of Indians, Pop. Sci. Mo., 32 (1888), p. 31.

[4] Garden and Forest, 9 (1896), p. 12.

species in use by the Japanese. Of these, *Lilium glehni* forms the chief
vegetable diet of the Ainu, an aboriginal tribe now confined to the
islands of Hokaido, but *L. tigrinum* and *L. concolor pulchellum* are the
two species most commonly cultivated as articles of food. Penhallow[1]
notes the use of *L. cordifolium*, and in a comparatively recent work[2] *L.
tigrinum*, *L. auratum*, and *L. elegans* are included among the plants cultivated for their edible roots. Bretschneider[3] also notes the occurrence
of several species of lilies with edible roots at Peking, one of which is
L. tigrinum. Davy[4] notes the use of the bulbs of *L. japonicum brownii*,
L. cordifolium, *L. tigrinum*, *L. concolor pulchellum*, and *L. glehni*, and
the dried flowers of *L. bulbiferum* and *Hemerocallis graminea* as food by
the Chinese and Japanese. (See also p. 44.) The author also found
the bulbs of *L. parvum* in use by the Washoe Indians of Nevada and
those of *L. pardalinum* in use by Indians of northern California.

From the early part of December to the latter part of August there
are found in the Chinese markets of San Francisco the bulbs of a species of Lilium which greatly resemble those of the well-known *L. auratum*. These are sold at the rate of from 10 to 20 cents a pound. They
are all imported from Canton. The bulbs have proved to be identical
with the ones sold by our nurserymen under the name of *L. brownii*,
and this is apparently the only species sold by the Chinese merchants,
as a large number of bulbs purchased at different times and from different dealers have invariably yielded plants corresponding to this
species. Unfortunately the bulbs are often infested with mites which,
either primarily or secondarily, cause the death of the plant before it
perfects its flowers. From a collection of over 100 bulbs only 10 perfect
flowers were secured.

The name *L. brownii* seems to have been first published by Poitean,[5]
but has since been regarded as synonymous with *L. japonicum*[6] or has
been reduced to merely varietal significance. The latter disposition
has been adopted here as being the most desirable from a cultural
standpoint.

What seems to be the same species may also be obtained in a dry
form throughout the year, and both this and the fresh bulbs are known
under the name of "pak hop," and designated by the Chinese characters 百合.

In Table 5 the analyses of lily bulbs purchased in San Francisco are
reported. One of them was of fresh the other of dried bulbs. Two
analyses of Japanese-grown bulbs are also quoted for purposes of comparison.

[1] Amer. Nat., 16 (1882), p. 119.
[2] Useful Plants of Japan. Tokyo, 1895.
[3] Jour. China Branch Roy. Asiatic Soc., 15 (1880), p. 179.
[4] Erythea, 6 (1898), p. 26.
[5] Rev. Hort., 2. ser., 2 (1843–44), 496, quoted from Index Kewensis, 3 (1894), p. 81.
[6] Baker, Revision of Tulipeae, 1875.

TABLE 5.—*Composition of lily bulbs.*

	Water.	Protein.	Albuminoids.	Amids (by difference).	Fat.	Starch.	Cane sugar.	Reducing sugars.	Crude fiber.	Ash.	Undetermined.
Lilium japonicum brownii (dried bulbs):	Per ct.	Per ct	Per ct.	Per ct.	P. ct.	Per ct.	Per ct.	Per ct.	P. ct.	P. ct.	P. ct.
Original material....	10.16	5.57	5.00	0.57	0.37	62.65	2.84	0.00	1.64	2.68	14.09
Water-free substance		6.20	5.56	.64	.41	69.73	3.16	.00	1.82	2.98	15.68
Lilium japonicum brownii (fresh bulbs):											
Original material....	66.72	2.33	1.55	.83	.59	17.74	4.16	.00	.75	1.24	6.42
Water-free substance		7.01	4.50	2.51	1.78	53.40	12.51	.00	2.25	3.74	19.32
Lilium tigrinum: a											
Original material....	71.46	4.51	....	....	.24	c 21.60			1.04	1 15	......
Water-free substance	15.79	....	....	.84	c 75.70				3.64	4.03	......
Lilium sp. ("Yuri"): b											
Original material....	69.63	3.40	....	....	.11	d19.10	....	....	1.42	1.35	......
Water-free substance		11.19	....	....	.05	62.83	....	....	4.66	4.44	......

a Reported by Kellner, quoted from König, Chemie der menschlichen Nahrungs- und Genussmittel, 3. ed., I, p. 794.
b Reported by Nagai and Murai, quoted from König, Chemie der menschlichen Nahrungs- und Genussmittel, 3. ed., I, p. 795.
c Carbohydrates by difference.
d In addition the authors report in the fresh substance 0.62 per cent of glucose, 2.44 per cent of pectose, and 1.92 per cent of dextrin.

Wide differences are to be noted between the composition of the two samples of bulbs purchased in San Francisco and the Japanese bulbs. This is not surprising when it is remembered that the bulbs were of different species and were grown in different countries.

The protein content is much smaller in the samples of *L. brownii* than in the others. In all the analyses the amount of protein is somewhat above the average amount found in the potato. The percentage of albuminoids is noticeably greater than in the potato. Starch forms by far the most important constituent of the bulbs and is present in sufficient amount to warrant the belief that they have a high nutritive value (Pl. III, fig. 3). The analysis quoted by Nitobe is interesting as giving somewhat definite information concerning the distribution of the carbohydrates, though he gives no statement as to the methods used in obtaining these results. Without them an accurate knowledge of the compounds present is scarcely possible. In the samples analyzed in this laboratory no cane sugar was found, though the presence of an abundance of mucilaginous or pectose-like substance was easily recognized.

The Chinese regard lily bulbs more as a delicacy than as a standard article of diet, and the customary price is considerably above that of other vegetables in common use by them. A Japanese friend informed the author that they are regarded by the Japanese as an especially desirable food for invalids and convalescents. When used for this purpose the bulbs are only slightly cooked and are eaten after the addition of sugar.

The bulbs sold in San Francisco, as far as the author's observations go, are nearly devoid of the bitter principle which is reported to occur in several species of Lilium. When simply boiled, they form a palatable food, and Americans might doubtless soon become accustomed to their use. The cultural conditions favorable to the production of *L. brownii* or of some of the other edible species are not difficult to find in our own country, though it is very doubtful whether they can be grown as cheaply as our other commonly cultivated vegetables. One valuable feature of the bulbs is the ease with which they may be dried, the resulting product being quite as acceptable as the fresh bulbs. The value of lilies as ornamental plants under present conditions would doubtless prevent their extended use as food in this country.

CHINESE SWEET POTATOES.

Two peculiar varieties of the sweet potato are on sale in the Chinese quarter of San Francisco and are noteworthy from the fact that they are the only edible roots found there which are familiar in American homes. One of these is a yellow variety with somewhat angular-pointed roots. It was at first supposed to be a true yam, or Dioscorea, but by growing a plant from one of the tubers it was shown that they were the product of *Ipomœa batatas* (番薯). This plant produced the lobed leaves characteristic of the yam-like varieties of the sweet potato, but it was impossible to connect it with any of the American varieties of which descriptions were accessible.[1]

A second red variety of *I. batatas* with rounded ends, which is also different from our commonly cultivated forms, is largely used by the Chinese. A plant grown from one of these roots produced the cordate leaves characteristic of a large number of varieties of this species. In Table 6 is shown the composition of the two sorts of Chinese sweet potatoes. The composition of the average of a number of American varieties is quoted for purposes of comparison.

TABLE 6.—*Composition of sweet potatoes.*

	Water.	Protein.	Albuminoids.	Amids (by difference).	Fat.	Starch.	Cane sugar.	Reducing sugars.	Crude fiber.	Ash.	Undetermined.
Ipomœa batatas (roots with pointed ends):	*Per ct.*	*Per ct.*	*Per ct.*	*Per ct.*	*Per ct.*	*Per ct.*	*Per ct.*	*Per ct.*	*Per ct.*	*Per ct.*	*Per ct.*
Original material	73.44	0.78	0.77	0.01	0.25	14.65	1.71	4.07	1.02	0.85	3.22
Water-free substance		2.95	2.90	.05	.95	55.16	6.43	15.34	3.85	3.20	12.12
Ipomœa batatas (roots with rounded ends):											
Original material	77.47	.73	.70	.03	.22	9.67	4.02	3.19	.93	1.00	2.76
Water-free substance		3.26	3.11	.15	.98	42.93	17.88	14.16	4.13	4.42	12.23
American sweet potatoes (average of 95 analyses): [a]											
Original material	69.0	1.80			.70	} 26.1			1.3	1.10	
Water-free substance		5.81			2.26	} 84.19			4.19	3.55	

[a] U. S. Dept. Agr. Office of Experiment Stations Bul. 28 (rev. ed.), p. 68.
[b] Carbohydrates by difference.

[1] For description of these, Texas Sta. Bul. 28, Georgia Sta. Bul. 30, and Louisiana Sta. Bul. 13 were consulted.

As shown by their composition neither of the Chinese sweet potatoes possesses any advantage over the ordinary American varieties. Both of these forms of sweet potato are said to be grown in California, but they are probably of Chinese origin. Apparently they possess no features which would render them worthy of introduction.

YAM BEAN.

The name "fan ko" is used by the Chinese of San Francisco to designate two different tuberous roots. One of these is globular or napiform in shape and varies from 1 to 2 decimeters in diameter. It is covered with a thick, yellow, stringy bark, which readily peels off and leaves a white fleshy interior of firm consistency and sweet taste. It agrees well with the root described by Henry[1] under the Chinese name given above, and the characters there used for this name (番葛) are the same as those used in San Francisco. Henry surmises that this is the root of *Pachyrhizus angulatus* (Pl. IV, fig. 2), a leguminous vine which is a native of Central America and is now widely diffused throughout the Tropics, although he was unable to secure definite proof of this statement. As it was found impossible to cause these roots to grow, a letter of inquiry was addressed by the author to Mr. Charles R. Ford, director of the Botanic Gardens at Hongkong, regarding them. In reply Mr. Ford stated that he had proved by actual experiment that the roots described by Henry were the product of *P. angulatus*.

The current descriptions of this plant give but little information regarding the character of the root. It is stated by one authority[2] that there is a cultivated form of the yam bean properly referred to *Dolichos tuberosus*, but later transferred to the genus Pachyrhizus by Sprengel, which differs from *P. angulatus* in the more rounded, nondentate leaves, the white and not violet flowers, and the larger pods. The opinion is also advanced that this form is specifically distinct, and that the name *P. tuberosus* should be adopted provisionally. Forbes and Hemsley[3] recognize only *P. angulatus* from China, and this is said to be in cultivation. Fawcett[4] states that the root of *P. tuberosus* is formed of a number of simple cord-like fibers, several feet in length, stretching under the surface of the ground, bearing in their course a succession of tubers. This description is in entire accord with the roots found in San Francisco. These have been called in this bulletin *P. angulatus* without, however, expressing any opinion whether the form known as *P. tuberosus* is distinct.

The composition of *P. angulatus* purchased in San Francisco and of *P. angulatus* and *P. tuberosus* grown in British Guiana, analyzed by Harrison and Jenman, is shown in the following table:

[1] Notes on the Economic Botany of China, p. 58. Shanghai, 1893.
[2] Kew Misc. Bul., 1889, pp. 17, 62, 121.
[3] Index Floræ Sinensis, 1887, p. 194.
[4] Economic Products of Jamaica, 1891, p. 37.

TABLE 7.—*Composition of yam bean roots.*

	Water.	Protein.	Albuminoids.	Amids (hydrat for sugar).	Fat.	Starch.	Cane sugar.	Reducing sugars.	Crude fiber.	Ash.	Undetermined.
	Per ct.	Per ct.	Per ct.	Per ct.	P. ct.	Per ct.	Per ct.	Per ct.	P. ct.	P. ct.	P. ct.
Pachyrhizus angulatus:											
Original material....	78.09	2.18	1.44	0.73	0.18	8.45	3.71	1.84	1.43	0.80	3.31
Water-free substance		9.84	6.59	3.35	.80	38.58	16.95	8.41	6.53	3.65	15.14
Pachyrhizus angulatus: a											
Original material....	85.70	......	1.04	......	.51	b 2.03	......	1.38	.71	1.37	......
Water-free substance		......	7.27	......	3.56	14.19	......	9.65	4.96	9.58	......
Pachyrhizus tuberosus: a											
Original material....	82.25	......	1.05	......	.30	c 8.46	1.29	.26	.66	1.84	......
Water-free substance		......	5.91	......	1.69	47.68	7.27	1.46	3.72	10.37	......

a Reported by Harrison and Jenman, Rpt. Agr. Work Bot. Gard. British Guiana, 1891-92, pp. 70, 71.
b In addition, the authors report in the fresh substance 0.17 per cent resin, 4.30 per cent digestible fiber, and 2.79 per cent gums, pectose, etc.
c In addition, the authors report in the fresh substance 0.15 per cent resin, 2.14 per cent digestible fiber, and 1.62 per cent gums, pectose, etc.

The analysis of the roots shows that they contain an abundance of nutritive materials. The protein content is also reasonably high, and viewed in the light of the analysis alone the roots might be pronounced a valuable food.

The roots also contain considerable cane sugar and a large amount of starch. The starch is present in nearly spherical grains which vary much in size. Compound grains of three or more individuals are not uncommon. The hilum is nearly central and is not elongated as in the case of most leguminous starches. No indication of stratification rings could be detected, and polarized light and a selenite plate failed to show a play of colors.

The varieties from British Guiana contain much more fat and ash than the Chinese specimens. The method of reporting the carbohydrates, however, differs from that used by the author, and the results are scarcely comparable. Judged by these analyses the Chinese-grown roots are superior from a dietetic point of view to those from South America.

Harrison and Jenman[1] note the presence of a poisonous resin in the tubers and seeds of *Pachyrhizus tuberosus* and a smaller amount in the seeds of *P. angulatus*. This resin was found to be a very active fish poison.

Long continued boiling of these roots failed to render them even reasonably tender, and left them with a sweet but otherwise insipid taste. The flesh is too tough to admit of being converted into a satisfactory esculent when judged by American tastes, though it has been stated by many authorities that the roots are used in this manner by the Chinese. Macfadyen[2] speaks in the highest terms of the edible qualities of the roots, and Dr. Trimen, of Ceylon,[3] found that both the roots and the green pods formed a valuable vegetable, but the more

[1] Rept. Agr. Work Bot. Gard. British Guiana, 1891-92, p. 70.
[2] Quoted by Fawcett. Loc. cit.
[3] Kew Misc. Bul., 1889, p. 17.

recent report of Harrison and Jenman[1] (British Guiana), does not praise the edible qualities of the root very highly. Their principal use by the Chinese, however, is for the preparation of a starch which is said to be of a superior quality.

As far as could be learned the Chinese obtain their comparatively large supply of roots entirely from Canton.

The plant requires a warm climate for its development, and probably there are but few localities in the United States in which it could be successfully grown.

A second root, for which the same Chinese name is used, but which is imported to a limited extent only, is often 1 meter in length, with a diameter of only about 12 centimeters. It also has a tough, stringy bark like the former species, and is still more tough and fibrous within than the former roots. There is but little doubt that this is the root of *Pueraria thunbergiana* (Pl. II, fig. 8), a leguminous vine which is a native of China and Japan and greatly resembles the former species. Specimens of this plant have been growing at Berkeley, Cal., for several years, but do not produce roots similar to those sold by the Chinese. This root, designated by the characters 粉葛, is so manifestly unfitted for use as an esculent that the author has not thought it worthy of an analysis. No satisfactory information could be obtained from the Chinese regarding their use of it. The roots are said to be used in China and Japan for the production of starch, and a fiber is also prepared from the stalks of the plant. As an ornamental vine the plants have many valuable features.

CASSAVA, OR MANIOC.

The characters here used 參茨 might be freely translated "the ginseng vegetable," and is not found in the works on Chinese botany or the Chinese-English dictionaries. It is used by the Chinese in San Francisco to designate a peculiar root which is about 5 decimeters in length and 1 in diameter and often weighs as much as 500 grams. The roots taper gradually at both ends and are covered with a gray bark about 8 millimeters in thickness. Within this is found a white sweet flesh which is traversed by a series of fine fibro-vascular bundles. Though it has been found impossible to grow plants from these roots, there is little doubt that they are the product of either *Manihot utilissima* (Pl. IV, fig. 4), the tapioca plant, or *M. aipi*, the sweet cassava—probably of the former species. The use of the above characters, meaning the ginseng plant, as a means of designating the roots evidently alludes to the habit of the plant itself, as both species of manioc present certain superficial resemblances to the *Panax ginseng* although altogether different botanically. The roots on sale in San Francisco are imported from China, and their principal use is in the preparation of a starch to which medicinal properties are attributed, although probably only from the fancied resemblance of the plant to the true ginseng. The author was also informed that the roots themselves were eaten after long-continued boiling.

The Division of Chemistry of this Department has published a bulletin[1] on sweet cassava, its culture, properties, and uses. This is based in part on the experience of Florida agriculturists.

The Florida Station has experimented for a number of years with cassava, and has recently published a number of results.[2] The plant has been very successfully grown. The bulletin describes methods of culture, the use of the roots for fattening pigs and steers, and as a food for horses and cows. The roots are also used for the manufacture of starch. In addition, a number of receipts are given for making puddings, fritters, and batter cakes from the freshly grated root. The cassava grown in Florida does not possess poisonous properties.

The composition of manioc, manioc flour, and cassava was also recently reported by Bóname.[3]

An analysis of one of the roots purchased in San Francisco, including the bark, is here reported. Other analyses are quoted for purposes of comparison:

TABLE 8.—*Composition of cassava roots.*

	Water.	Protein.	Albuminoids.	Amids (by difference).	Fat.	Starch.	Cane-sugar.	Reducing sugars.	Crude-fiber.	Ash.	Undetermined.
Manihot utilissima[c]	Per ct.	Per ct	Per ct.	Per ct.	P. ct.	Per ct.	Per ct.	Per ct	P. ct.	P. ct.	P. ct.
Original material...	80.72	1.58	1.43	0.15	0.17	12.01	0.76	1.50	0.85	0.63	1.69
Water-free substance	......	8.24	7.40	.80	.88	62.28	3.95	8.25	4.43	3.28	8.71
Manihot dipi;a											
Original material....	64.30	.64	......	......	.17	30.98	......	......	.88	.51	5.52
Water-free substance	......	1.66	......	......	.44	80.06	......	......	2.26	1.31	14.27
Manihot utilissima;b											
Original material....	67.65	1.17	......	......	.40	23.10	......	......	1.50	.65	c5.53
Water-free substance	......	3.62	......	......	1.24	71.39	......	......	4.63	2.01	e17.08
Cassava; d											
Original material....	66.69	.85	......	......	.18	e29.84	......	......	1.68	.74	......
Water-free substance	......	2.55	......	......	.54	e89.65	......	......	5.04	2.22	......

a Reported by Wiley, U. S. Dept. Agr., Division of Chemistry Bul. 44, p. 9.
b Reported by Payen, Compt. Rend. Acad. Sci. Paris, 44 (1857), p. 401.
c Sugars, gums, etc.
d Reported by Stockbridge, Florida Sta. Bul. 49, p. 39.
e Carbohydrates by difference.

It will be seen that these roots are especially rich in starch, and contain considerable amounts of both cane sugar and reducing sugar. The fat content is low and the protein content somewhat greater than that of the average potato. Nearly 90 per cent of the protein is true albuminoid. It is well known that hydrocyanic acid occurs in most species of manioc, especially of *M. utilissima*, but the temperature at

[1] U. S. Dept. Agr., Division of Chemistry Bul. 44.
[2] Florida Sta. Bul. 49.
[3] Rap. An. Sta. Agron. [Mauritius], 1897, p. 57.

which the above sample was dried previous to analysis should have driven off the greater part of this; and even if present, it is scarcely probable that it would have materially increased the figures for either crude protein or albuminoids on account of the method used for these determinations.

The roots sold in San Francisco evidently differ greatly in composition from those analyzed by Wiley,[1] the ash being much greater, the fat much less, the protein nearly four times as great, and the starch and sugars somewhat less. The analysis of roots of *M. utilissima* quoted from Payen[2] more nearly approaches that of the San Francisco sample.

The starch grains found in the root purchased in San Francisco differ from those commonly figured in illustrating the starch of the sago, and also differ from the grains of *M. aïpi* as given by Wiley.[1] They are commonly oval or oblong and measure on the average 25 by 17.5 μ. The hilum is eccentric and the rings prominent.

Judged by the results of the analysis alone, if one has in view the utilization of the plant as a vegetable, the Chinese-grown roots appear to possess decided advantages. If they are the product of *M. utilissima*, however, the presence of hydrocyanic acid in amounts sufficient to render their continued use objectionable is to be feared. If the production of starch or glucose is the object desired, the roots of *M. aïpi* would be the more desirable of the two species, judged by the analyses reported above.

GREEN VEGETABLES AND CUCURBITS.

GREEN VEGETABLES.

The peculiar forms of cabbage and mustard used by the Chinese residents of our Eastern cities have already been fully described by Bailey.[3] Most of the forms there discussed are found on sale in San Francisco, and but little need be added to what he has written regarding their botanical characters.

Brassica pe-tsai, the variety of cabbage known as "pe-tsai," designated by the Chinese characters 白菜, appears to be more largely used and more highly esteemed than any other. This consists of a loosely compacted cluster of tender white leaves with greatly thickened midribs, the whole resembling a head of lettuce rather than a true cabbage. The central axis of the plant, however, extends to about one-half the length of the head. This is the "shantung cabbage," which, though long ago brought to the attention of European seedsmen and gardeners, seems never to have been distributed widely enough to admit of an accurate determination of its merits. The reports of Bailey and others[4]

[1] U. S. Dept. Agr., Division of Chemistry Bul. 44.
[2] Compt. Rend. Acad. Sci. Paris, 44 (1857), p. 401.
[3] New York Cornell Sta. Bul. 67, pp. 178–191.
[4] Kew Misc. Bul., 1888, p. 137; 1893, p. 344.

28

indicate that it has many valuable qualities and might be advantageously brought into more general use.

Forbes and Hemsley [1] refer the plant specifically to *B. campestris;* Smith,[2] Bretschneider,[3] and apparently most writers on Chinese botany refer it to *B. chinensis;* Loureiro,[4] however, regards it as a form of his *Sinapis pekinensis,* to which he ultimately gave the varietal name of "pe-tsai," and Bailey has raised this name to a specific one.

There are many varieties of "pe-tsai." The one sold in San Francisco differs from that figured by Bailey in the more densely compacted heads and in the greater spread of the leaf blades from the very base of the midrib. One head weighed 750 grams. When tested, this plant seemed to the author an entirely satisfactory vegetable.

The composition of "pe tsai" is shown in Table 9, p. 32. The analysis there reported shows that this vegetable does not differ greatly as regards food values from our commonly cultivated cabbages. The water content is somewhat greater than the average of the figures for American-grown samples. The protein content is also somewhat lower, but nearly two-thirds of it is of an albuminoid character. The presence of about 30 per cent. of reducing sugar, and also of considerable amounts of starch and small amounts of the cane members of the sugar group, is interesting, but corresponding determinations for other forms of cabbages were not found.

B. chinensis, another largely used cruciferous plant, is sold under the name of "pak-tsai," though the same Chinese character is used to designate it as is used for the former vegetable. The samples found in the San Francisco market are made up into bundles, and consist of either the young and tender leaves or the blanched petioles of the leaves or the stalks derived from the central axis of the plant itself. The latter sometimes show the yellow mustard-like flowers. Bailey designates this plant, though with some hesitation, by the above name, but Forbes and Hemsley include it under *B. campestris.*

The composition of "pak-tsai" is shown in Table 9, p. 32. In composition this vegetable shows nearly the same features as the one previously described except in the case of reducing sugars, which form only about 10 per cent of the dry material. The crude protein is also somewhat lower, but the figures for albuminoids are practically the same. On the whole, as regards composition the vegetable is only slightly inferior to that previously described, and seems worthy of a general trial.

B. juncea.—Still a third cruciferous vegetable is found on sale throughout almost the entire year in San Francisco. It consists of green stalks often 50 centimeters in length, which are usually derived from

[1] Index Florae Sinensis, 1886, p. 46.
[2] Contributions Toward the Materia Medica and Natural History of China, p. 42. Shanghai, 1871.
[3] Jour. China Branch Roy. Asiatic Soc., 15 (1880), pp. 31, 121.
[4] Flora Cochinensis, 1790, p. 400.

the petioles only of *B. juncea*, though early in the season the entire leaf is used. This vegetable is known as "kiai tsai," designated by the Chinese characters 芥菜, and is probably a form of *B. juncea*. Bretschneider[1] reports that several varieties of *Sinapis juncea*, or *B. juncea*, are included under the name "kiai tsai" at Peking, and gives the same character as is used in San Francisco for this vegetable. No analysis of "kiai tsai" was made in this laboratory, and none seems to have been reported by other investigators. This species often is sold in a dried form, and is also preserved in brine and used in the same way as sauerkraut.

Amarantus sp.—The young seedling plants of Amarantus are in common use as a pot herb among the Chinese of the Pacific coast. Both our native species and a form which is said to be regularly cultivated are in use. The plants found in market are too immature to admit of specific identification. This vegetable goes under the name of "in tsai," designated by the characters 莧菜, which Bretschneider[2] says is generally applied to several species of Amarantus. Bailey[3] describes a form of *A. gangeticus* used by the Chinese as a pot herb. Smith[4] speaks of the use of *A. oleraceus* and Forbes and Hemsley[5] state that *A. blitum*, *A. caudatus*, *A. gangeticus*, and *A. paniculata* are commonly cultivated as vegetables in China. Several varieties of indigenous Amarantus and the closely related Chenopodium are eaten to a limited extent in the United States. Their value has been pointed out by Coville. [6]

The composition of Amarantus purchased in San Francisco is shown in Table 9, p. 32. The analysis here recorded presents no unusual features save in the high content of crude protein, in which feature it surpasses the cruciferous vegetables mentioned above.

Solanum melongena.—A peculiar variety of the eggplant, known in San Francisco as "pak ke," designated by the characters 白茄, is one of the noticeable features of the Chinese vegetable stands during the summer months. This has a perfectly smooth white skin and is from 9 to 24 centimeters long and about 10 in diameter. Osbeck, Loureiro, and most of the early writers on Chinese botany speak of the cultivation of the eggplant. The composition of this eggplant is shown in Table 9, p. 32. Apparently but few analyses of this vegetable have been made, but the results obtained from the analysis of the Chinese variety in this laboratory agree quite closely with that of an American-grown specimen.[7] The percentage of sugars and starch is apparently quite large, though the real nature of the carbohydrates is uncertain since

[1] Jour. China Branch Roy. Asiatic Soc., 15 (1880), p. 92.
[2] Ibid., p. 168.
[3] New York Cornell Sta. Bul. 67, p. 199.
[4] Contributions Toward the Materia Medica and Natural History of China, p. 12. Shanghai, 1871.
[5] Index Florae Sinensis, 1891, p. 319.
[6] U. S. Dept. Agr., Yearbook 1895, p. 210.
[7] U. S. Dept. Agr., Office of Experiment Stations Bul. 28, p. 38.

they were determined by difference. Starch, however, as shown by qualitative tests in this laboratory, is certainly present in considerable amounts.

CUCURBITS.

The fruits of several species of Cucurbitaceæ are extensively used by the Chinese. Bailey[1] has already given an account of the cultural features and uses of four species, all of which were found on sale in San Francisco and are there used in large quantities. Further information concerning these vegetables is given by Lauder[2] and by Georgeson.[3]

Momordica charantia, the "fu kwa" or "khu qua" of Bretschneider, or the "la qua" of Bailey, is designated by the Chinese characters 苦瓜. This vegetable is noted in nearly all works descriptive of the botany of China, where it is largely cultivated and to which country it is probably indigenous. It is also sometimes called "lai kua," or the leprosy gourd. Its immature fruit is largely used throughout the Tropics as a condiment in the preparation of curries, etc. By the Chinese it is used especially in the preparation of salads, etc. The closely related *M. balsamina* is already somewhat widely known in America as an ornamental vine, and the seeds of both species are to be obtained from American seedsmen. The shuttle-shaped green fruits of *M. charantia* are about 2 decimeters in length, covered with rows of wart-like tubercles, and are borne on wiry stems of about the same length. Like all the other green vegetables described above, they are grown in large quantities in the Chinese gardens along the Sacramento River.

The *M. charantia* can scarcely be regarded as a food plant in the narrower sense of that term, as the principal value of its fruits depends upon the flavor they impart to other preparations, rather than on the amount of nutriment they contain. Nevertheless, large amounts of the fruits are eaten, and they must be considered in studying the dietary of the Chinese. The analysis here reported (Table 9, p. 32) shows that the fruit is not lacking in nutritive constituents, though the exact nature of the compounds represented could not be determined without a more extended investigation. The crude fat in this case apparently consisted almost wholly of chlorophyll. Of the 18 per cent of nitrogenous constituents, about two-thirds was of an albuminoid nature. Reducing sugars, or at least soluble carbohydrates having reducing power, were found to be present in considerable quantities, and true starch was also present in relatively large amounts.

Luffa acutangula, is known as "sz kwa," and both this and the following species are designated by the Chinese characters 絲瓜. This is a native of tropical Asia, though naturalized in certain regions of America. The name above given is synonymous with *L. fœtida* and

[1] New York Cornell Sta. Bul. 67, pp. 191–196.
[2] Garden and Forest, 1 (1888), p. 483.
[3] Amer. Garden., 13 (1892), p. 526.

with *Cucumis acutangulus*. The plant produces a green, pronouncedly ten-ribbed fruit, about 3 decimeters in length, which is obtusely terminated at the farther end, but tapers gradually to the point of union with the peduncle. In uses and culinary qualities it much resembles the common cucumber. In San Francisco, it is said, the Chinese use it for thickening soups. Bailey[1] states that it is eaten raw, and also cooked in the same manner as we use squashes. As shown by analysis (Table 9, p. 32), it does not differ greatly from the species previously described, the chief difference being in the lower percentage of nitrogenous substances and in the greater content of carbohydrates, especially of reducing sugars.

Luffa cylindrica is called by the Chinese "po kua." This is the "sua kwa" of Bailey.[2] This plant is also widely cultivated throughout the Tropics, but its native country is unknown. It is synonymous with *L. aegyptiaca*, *L. petiola*, and *Momordica cylindrica*. It is reported from China by most of the naturalists who visited that country in early days, as well as by the more recent botanists. Its slender, cylindrical, crooked, yellowish-green fruits, which are often 6 decimeters long, are used in the immature condition in the same manner as we use squashes. The interior of the mature fruit is filled with a fibrous mass, which when dried forms a useful household article. Hence the plant is often known as the "dish-cloth" or "towel gourd," and is more often grown for its fibrous interior than as a food. In San Francisco, however, it is largely used for the latter purpose, being prepared in much the same manner as squashes are treated in American households. As shown by analysis (Table 9, p. 32), it is on the whole about equivalent in food value to the previously-described species, the protein content being somewhat less, while the percentage of reducing sugars and starch is somewhat greater. As shown by comparison, it is rather inferior to the common varieties of squash, and it can scarcely be regarded as more than a passable substitute for them.

Benincasa cerifera.—This plant is known as "zit kwa," and is designated by the characters 節瓜. It is the same as the "tung kua" of Bretschneider,[3] designated by the characters 冬瓜. Under the name of "Chinese preserving melon," it is already somewhat known in the United States. It is a tropical vine, long cultivated in China, Japan, India, and Africa, where it is often met with growing spontaneously, though its native country is unknown. The young green hairy fruits when about 20 centimeters in length, are used in the same manner as we use the squash. The mature fruits often attain a weight of 12 kilograms, are perfectly smooth, and covered with a white wax. With the exception of a small amount of pulpy substance which is filled with the white seeds, the interior is made up of a white solid flesh. The

[1] New York Cornell Sta. Bul. 67, p. 196.

[2] Ibid., p. 195.

[3] Jour. China Branch Roy. Asiatic Soc., 15 (1890-91), p. 153.

mature fruit is used by the Chinese in the preparation of confections and is also said to be used as a vegetable. The analysis reported below presents no especially interesting features save the high percentage of reducing sugars.

The composition of the green vegetables and cucurbits discussed above is shown in the following table:

TABLE 9.—*Composition of green vegetables and cucurbits.*

	Water.	Protein.	Albuminoids.	Amids (by dif. ference).	Fat.	Starch.	Cane sugar.	Reducing sugars.	Crude fiber.	Ash.	Undetermined.
GREEN VEGETABLES.											
	Per ct.	Per ct.	Per ct.	Per ct.	P. ct.	Per ct.	P. ct.	Per ct.	Per ct.	P. ct.	P. ct.
Brassica petsai:											
Original material....	95.74	1.19	0.48	0.71	0.15	0.31	0.09	1.20	0.52	0.56	0.14
Water-free substance	28.07	11.96	16.71	3.57	7.19	2.11	30.25	12.16	13.28	3.38	
Brassica chinensis:											
Original material....	96.55	.78	.41	.37	.10	.31	.09	.37	.46	.65	.80
Water-free substance	21.96	11.43	10.53	2.82	8.61	2.45	10.45	12.86	18.33	22.52	
Amarantus sp.:											
Original material....	91.52	2.61	1.67	.94	.36	.50			.92	1.56	2.55
Water-free substance	30.80	19.68	11.12	4.26	5.82			10.81	18.31	30.00	
Solanum melongena:											
Original material...	89.62	1.38	1.08	.29	.30	1.57	.63	1.31	1.54	.69	2.95
Water-free substance	13.25	10.42	2.83	2.91	15.11	6.10	12.66	14.88	6.69	28.39	
CUCURBITS.											
Momordica charantia:											
Original material....	93.61	1.18	.79	.39	.20	.67	.06	.60	1.07	.31	2.28
Water-free substance	18.48	12.31	6.18	3.19	10.56	.71	9.36	16.72	5.25	35.69	
Luffa cylindrica:											
Original material....	94.66	.51	.38	.13	.19	1.04	.12	2.15	.46	.41	.45
Water-free substance	9.57	7.07	2.50	3.72	19.52	2.18	40.29	8.58	7.65	8.49	
Luffa acutangula:											
Original material....	94.90	.68	.54	.14	.24	.36	.10	1.57	.72	.43	1.00
Water-free substance	13.39	10.68	2.71	4.70	7.03	1.95	30.86	14.03	8.43	19.61	
Benincasa cerifera:											
Original material....	96.24	.50			.16	.31	.07	.90	.57	.35	.88
Water-free substance	13.27			4.34	8.29	1.74	24.19	15.19	9.48	23.49	

SEEDS AND GRAINS.

SOY BEANS.

Leguminous seeds and certain preparations made from them have always formed an important part of the largely vegetarian diet of the Chinese and Japanese. Of the legumes the soy bean, *Glycine hispida* (*Soja hispida*), is the most important. The soy bean has long been known in Europe. Kaempfer[1] was perhaps one of the first Europeans to describe it. This plant has been cultivated many years in Europe, and is coming to be quite extensively grown in the United States, largely for use as a forage plant. The soy bean has been treated of in a previous publication[2] of this Department. A large number of varieties of the soy bean are in cultivation in China and Japan, but only two were found in the Chinese markets in San Francisco, a yellow and a black variety. Aside from a difference in color, the two forms

[1] Amenitatum Exoticarum, p. 837. Lemgoviae, H. W. Meyer, 1712.
[2] U. S. Dept. Agr., Farmers' Bulletin 58.

U. S. Dept. of Agr., Bul. 58. Office of Expt. Stations.

PLATE VI.

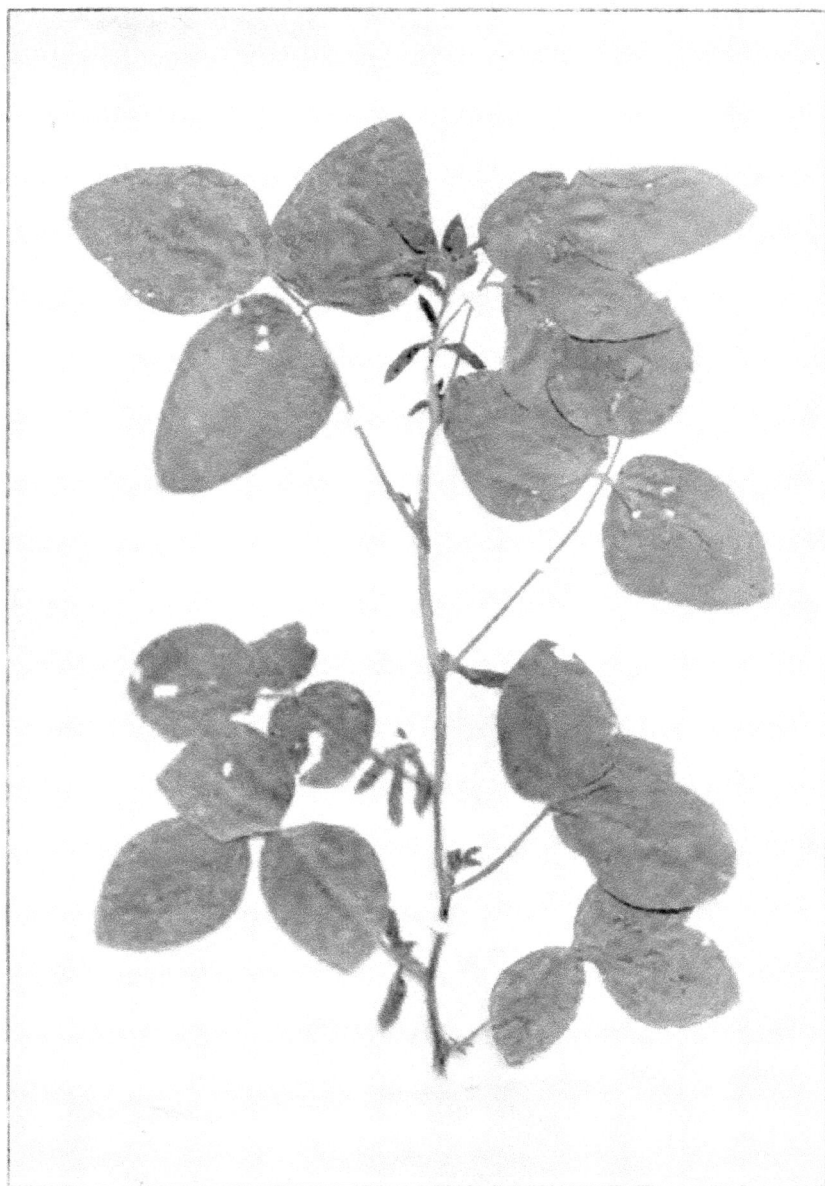

UPPER PORTION OF A PLANT OF THE BLACK SOY BEAN.

apparently do not differ materially from each other. The yellow variety is known as "wong tau," and is designated by the characters 黄豆, while the black is known as "hak tau," and is designated by the characters 黑豆. The soy bean resembles a pea rather than a bean, although the botanical characteristics of the plant indicate that it is very different from any of our cultivated peas or beans.

Both varieties obtained from the Chinese market in San Francisco grew readily in Berkeley, attaining a height of about 3 feet, and in spite of a very dry season produced an abundant crop of seeds. The appearance of two of these plants at different stages of growth is shown in Pls. VI and VII. The composition of the seeds of the two varieties is shown in Table 10, the average composition of American-grown soy beans being quoted also for purposes of comparison.

TABLE 10.—*Composition of soy beans.*

	Water.	Protein.	Albuminoids.	Amids(by difference).	Fat.	Starch.	Cane sugar.	Reducing sugars.	Crude fiber.	Ash.	Undetermined.
	Per ct.	*Per ct.*	*Per ct.*	*Per ct.*	*Per ct.*	*Per ct.*	*Per ct.*	*P. ct.*	*P. ct.*	*P. ct.*	*P. ct.*
Glycine hispida (black):											
Original material....	8.25	36.35	34.63	1.72	17.22	6.80	7.38	0.00	5.25	4.77	13.98
Water-free substance		39.62	37.74	1.82	18.77	7.41	8.04	0.00	5.72	5.20	15.25
Glycine hispida (yellow):											
Original material....	8.33	36.00	35.54	.46	17.87	5.87	6.55	0.00	5.43	4.75	15.20
Water-free substance		39.27	38.77	.50	19.49	6.40	7.14	0.00	5.92	5.18	16.59
Glycine hispida (average of 8 analyses):a											
Original material....	10.80	35.98	......	......	16.85		b28.89		4.79	4.69	......
Water-free substance		38.1	......	......	19.00		b32.2		5.4	5.3	......

a U. S. Dept. Agr., Office of Experiment Stations Bul. 11, p. 120.
b Carbohydrates by difference.

The analyses of these two Chinese varieties of soy beans, as well as of others which have been published, show that they contain an unusually large amount of valuable food constituents, especially protein and fat, the former constituting about 39 and the latter about 19 per cent of the total weight. As shown by the analysis made by the author, nearly all the protein is made up of albuminoids. The figures obtained by the author for cane sugar and starch indicate that a considerable amount of carbohydrates was present, although starch could not be detected by the ordinary test. No special attempt was made to separate the cane sugar.

There has been considerable difference of opinion among investigators regarding the presence of starch in soy beans. The weight of evidence, however, seems to indicate that, under certain conditions, starch in considerable amounts may be present. In 1880 Meissl and Böcker[1] investigated soy beans and reported the presence of starch. The amount of starch found was said to be less than 3 per cent. The

[1] Sitzber. Math. Naturw. Cl. Akad. Wiss. (Vienna), 87 (1883), pt. 1, pp. 372–391.

starch grains are described as being extremely small and as differing in form from the typical bean or pea starch. About the same time Hanausek[1] also reported starch in soy beans. It was found deposited where the surface of the cotyledons met, and according to the author, could not be detected by the usual chemical methods. The starch grains were embedded in fat and gave no color reaction after prolonged treatment with iodin. A more extended study of soy beans was reported by Harz.[2] According to this author, when the beans do not ripen thoroughly, or when allowed to ripen after the vines are cut, they may contain starch, some varieties being more likely to contain it than others. Inoyne[3] also was unable to find starch in the mature seeds.

The carbohydrates other than starch in soy beans have been studied by many investigators. Stingl and Morawski[4] report the presence of small quantities of dextrin in soy beans, and about 2 per cent of a mixture of different sugars which could be readily fermented. Their investigation was principally concerned with the diastatic ferment found in the beans. A number of tests on the quantity of the ferment are reported. La Vellois[5] reported the presence of 9 to 11 per cent of material soluble in alcohol, which did not reduce Fehling's solution. Tollens[6] calls this substance galactan. Schultz and Frankfurt[7] proved the presence of cane sugar in soy beans. Maxwell[8] identified paragalactan, an insoluble carbohydrate. A number of other investigations on the carbohydrates of soy beans have been reported.

In 1880 Meissl and Böcker[9] reported an extended investigation of the proteids of soy beans. According to these investigators the beans contained the following proteids: A so-called "casein" (27.6 per cent), albumen (0.5 per cent), a proteid precipitated by cupric oxid and potassium hydroxid (2.5 per cent), and, in addition, a very small amount of nonalbuminoid nitrogenous substance. Osborne and Campbell[10] have recently reported an extended study of the proteids of soy beans. The principal proteid found is called "glycinin." It is a globulin. According to these authors the beans also contain about 1.5 per cent of an albumin-like proteid, legumelin. A small amount of proteose was also found.

Meissl and Böcker[9] studied the fat of soy beans, which is said to contain no free fatty acid and to consist almost entirely of neutral trigly-

[1] Ztschr. Allg. Oester. Apoth. Ver.. 22 (1881), p. 474.
[2] Ztschr. Allg. Oester. Apoth. Ver., 25 (1885), p. 40; Landwirtschaftliche Samenkunde, p. 692. Berlin, P. Parey, 1885.
[3] Imp. Univ. Col. Agr. [Tokyo] Bul., Vol. 2, No. 4.
[4] Monatsh. Chem., 7 (1886), pp. 176-190.
[5] Compt. Rend Acad. Sci. Paris, 93 (1881), p. 281.
[6] Handbuch der Kohlenhydraten, vol. 1, p. 208. Breslau, E. Trewendt, 1888.
[7] Ber. Deut. Chem. Gesell., 27 (1894), p. 62.
[8] Amer. Chem. Jour., 12 (1890), p. 51-60.
[9] Loc. cit.
[10] Conn. State Sta. Rpt. 1897, pt. 4, pp. 371-382.

MATURE PLANT OF YELLOW SOY BEAN.

cerids. At a low temperature or on standing a long time the palmatin and stearin triglycerids were precipitated in crystalline form. The fat has a characteristic leguminous taste. When kept for two years it became thick, but was only slightly rancid. Its specific gravity was 0.89 at 15° C. One gram of soy-bean fat required 191.8 milligrams potassium hydroxid for saponification. The fat has also been examined by Stingl and Morawski[1] and by Roelofsen[2] with reference to its iodin absorption. The figures obtained for the latter constant were 121.3 and 138.8, respectively, and these and other facts indicate that the oil has poor drying properties.

Pellet[3] studied the ash of soy beans in considerable detail.

Soy beans are eaten to some extent when cooked in the same ways as other beans. The Chinese express an oil from them which is a standard article of commerce and is largely used in cooking. The principal use which they make of the soy beans, however, is in the preparation of a vegetable cheese, a kind of thick sauce, and other products. An account of the manufacture of miso and soy-bean sauce in the early part of the eighteenth century is given by Kaempfer.[4] According to Prinsen-Geerligs,[5] " tao hu," or bean cheese, is prepared from the seeds of the white variety of soy bean. These are allowed to soak for three hours in water, are then reduced to a thick paste, and the mass cooked. The cooked mass is strained through a coarse cloth. The filtrate consists of a milky-white liquid containing protein and fat. As soon as this becomes cool some material is added (for instance, crude salt containing magnesium chlorid), which precipitates the proteid material, the fat being inclosed in the coagulated mass. The coagulated material is pressed and kneaded into small cakes. The cakes may be dipped for a few moments into a saline solution of curcuma. Variations in the process give rise to a number of varieties of bean cheese. This is essentially the method used by the Chinese of San Francisco in the preparation of the bean cheese used by them. It is sold either in the form of the freshly precipitated curd or in the form of small square cakes obtained by compressing the former material. It is usually cooked in peanut oil before being eaten and, in the author's opinion, is a palatable food. A partial analysis of one of these cakes gave water 81.35 per cent, fat 5.19 per cent, and ash 0.80 per cent.

The filtrate from the cooked soy beans resembles milk, and, on heating, a skin, not unlike that formed on milk, rises to the surface of it.

A thick sauce called "tao yu," resembling the Japanese "shoju," is also prepared from the soy beans, as well as a thick condiment called "tao tjiung," similar to the Japanese "miso." According to Prinsen-Geerligs, these preparations have the following composition:

[1] Chem. Ztg., 10 (1886), p. 140.
[2] Amer. Chem. Jour., 16 (1894), p. 19.
[3] Compt. Rend. Acad. Sci. Paris, 90 (1890), pp. 1177–1180.
[4] Loc. cit.
[5] Chem. Ztg., 20 (1896), pp. 67–69.

Table 11.—*Composition of Chinese soy-bean preparations.*

	Water.	Protein.	Fat.	Carbohydrates.	Ash.	Undetermined.
	Per cent.	Per cent.	Per cent.	Per cent.	Per cent.	Per cent.
Tao hu (bean cheese)	76.15	13.15	7.09	1.40	2.21	....
Milk from boiled soy beans	93.10	3.13	1.89	....	0.51	....
Tao yu (soy sauce)	57.12	7.49	....	16.03	a 18.76	....
Tao ijung	62.86	12.67	1.21	b 13.78	6.71	2.77

a Including 17.11 sodium chlorid.   b Including 3.78 crude fiber.

The Japanese preparation made from soy beans, similar to those mentioned above, have been described by a number of investigators. A brief account of them has been given in a previous publication of this Department.[1] As shown by their composition. these soy-bean preparations contain a high percentage of nutrients. They are eaten in large quantities and form important articles of diet. A number of digestion experiments have been reported in which the Japanese preparations formed a considerable part of the diet, and on the basis of the results obtained the preparations are considered to be very well assimilated.

PHASEOLUS.

The seeds of two varieties of *Phaseolus mungo* are largely used by the Chinese in San Francisco. One of these, known as "luk tan," designated by the characters 綠豆, is a small green bean. The individual seeds weigh only about 0.044 gram. They are slightly flattened at the ends and have a rather long hilum. Plants were grown at Berkeley, Cal., from these seeds and yielded a good return, though the plant was not as prolific as the soy bean. The appearance of one of them is shown in Pl. VIII. The analyses in the table below show that these beans differ in composition but slightly from the commonly cultivated varieties of *P. vulgaris*.

The Chinese use these beans largely in the preparation of "bean sprouts," which are simply seeds that have been soaked in water and allowed to germinate till the young plants are several inches in length. This product is said to be used in making soap. It is difficult to imagine what advantage the young plants possess over the original seeds for such a purpose, unless they impart a different flavor. From an economic standpoint the process is wasteful, as it involves the transformation of albuminoids into various cleavage products and amido-compounds, whose nutritive value is thought to be but slight. Henry[2] refers to the preparation of a kind of vermicelli from this bean. but such a preparation was unknown in San Francisco.

The second form of Phaseolus in use is of a dull red color. rather larger than the previous variety. and more nearly spherical in shape. This is known as the "huang tan" or red bean, designated by the characters 紅豆. Kellner analyzed the seeds of a variety of *P. radiatus*

[1] U. S. Dept. Agr., Farmers' Bul. 58, Appendix.
[2] Notes on the Economic Botany of China, p. 13, Shanghai, 1893.

PLATE VIII.

PLANT OF *Phaseolus mungo.*

grown in Japan, which is very probably the same as the one here described, as the two species are by some authors regarded as identical. Harrison and Jenman report the analysis of mungo beans grown in British Guiana.

The composition of the red and green *P. mungo* is shown in Table 12. For purposes of comparison the composition of a green form of *P. mungo* reported by Church, *P. radiatus* reported by Kellner, and "mungo" beans reported by Harrison and Jenman, are also given.

TABLE 12.—*Composition of seed of Phaseolus spp.*

	Water.	Protein.	Albuminoids.	Amids (by difference).	Fat.	Starch.	Cane sugar.	Reducing sugars.	Crude fiber.	Ash.	Undetermined.
Phaseolus mungo (green)	Per ct.	Per ct.	Per ct.	Per ct.	Per ct.	Per ct.	Per ct.	P. ct.	P. ct.	P. ct.	P. ct.
Original material....	8.83	22.64	21.88	0.76	1.54	48.54	0.00	0.00	4.52	2.85	11.28
Water-free substance.......		24.83	24.00	.83	1.47	53.23	.00	.00	4.95	3.13	11.39
Phaseolus mungo (red):											
Original material....	10.47	21.06	18.19	2.87	.61	48.36	1.65	.00	5.02	3.22	9.61
Water-free substance.......		23.52	20.32	3.20	.68	54.02	1.84	.00	5.61	3.60	10.73
Phaseolus mungo (green) a											
Original material....	10.8	22.2	....	....	2.7	....	b54.1		5.8	4.4	....
Water-free substance.......		24.9	....	....	3.0	....	b60.7		6.5	4.9	....
Phaseolus radiatus (red) c											
Original material....	12.20	18.30	16.74	1.56	1.42	57.40	d2.02		6.05	2.60	....
Water-free substance.......		20.84	19.06	1.78	1.62	65.38	d2.34		6.89	2.96	....
"Mungo beans:" e											
Original material....	14.75	....	20.54	1.52	1.96	f6.71	3.65	.71	3.74	3.39	....
Water-free substance.......		....	24.09	1.78	2.29	7.87	4.28	.83	4.38	3.98	....

a Reported by Church, Food Grains of India. London, Chapman and Hall, 1886, p. 151.
b Carbohydrates by difference.
c Reported by Kellner, Landw. Vers. Stat., 30 (1884), p. 42.
d Determined by difference.
e Reported by Harrison and Jenman, Rpt. Agr. Work Bot. Gard. British Guiana, 1891-92, p. 73.
f The authors also report in the fresh substance 41.23 per cent digestible fiber and 1.80 per cent pectose, gums, etc.

Harrison and Jenman[1] also report analyses of a number of beans whose botanical names were not known. They were called red "mote," "Chinese bean," "kingto," and "octow."

The results of Kellner's analysis, as well as those secured at San Francisco (the red *Phaseolus mungo*), show that the two varieties have essentially the same composition. The analyses also show that about 90 per cent of the protein is in the form of albuminoids. Osborne[2] has studied the proteids of the white-podded adzuki bean, which is presumably the same variety as the one analyzed by Kellner. He found they consisted of phaseolin and a hitherto unknown globulin.

DOLICHOS.

Still another bean, which is largely used by the Chinese of San Francisco, is *Dolichos sesquipedalis*. These seeds are white, with a prominent black ring which surrounds the hilum. The beans weigh about 0.31 gram each. The skin covering the seed is especially tough, but

[1] Loc. cit.
[2] Jour. Amer. Chem. Soc., 19 (1897), p. 509.

can readily be removed after soaking in water. The beans are most largely used in the form of "bean sprouts" after the removal of this skin. They are known as "mi tau" (米豆); but this name and its accompanying characters have not been found in any of the books on Chinese botany now accessible to the author. The plant is said to be a native of South America, and was probably recently introduced into China. As shown below, the composition of the dry bean differs but slightly from that of our commonly cultivated variety of beans.

The green pods of this species are also largely used as a snap bean. These are from 50 to 70 centimeters in length, and contain from 10 to 16 seeds. They are more slender than most of our string beans, and slightly ridged along the middle of the two valves. This vegetable is known as "tou kok," and is cultivated by the Chinese along the Sacramento River, arriving in market early in July. It is said that it has come into use among the white residents of the central portion of California, and is considered a valuable variety of string bean. The plant is somewhat tender, and requires a long season for development, but Bailey[1] reports fair success with it in the eastern United States.

Table 13 shows the composition of the seeds and green pods of *D. sesquipedalis* as compared with that of *D. lablab*, cowpeas, and other beans.

TABLE 13.—*Composition of Dolichos and other beans.*

	Water.	Protein.	Albuminoids.	Amids (by difference).	Fat.	Starch.	Cane sugar.	Reducing sugars.	Crude fiber.	Ash.	Undetermined.
	Per ct.	Per ct.	Per ct.	P. ct.	P. ct.	Per ct.	Per ct.	Per ct.	Per ct.	P. ct.	P. ct.
Dolichos sesquipedalis (seed):											
Original material....	10.98	22.74	17.84	4.90	2.66	45.91	4.27	0.00	2.80	2.86	7.78
Water-free substance	25.54	20.04	5.50	2.99	51.57	4.80	.00	3.15	3.21	8.74	
Dolichos lablab (seed):a											
Original material....	12.1	24.4	....		1.5	b 57.8			1.2	3.4	....
Water-free substance	27.8	....		1.7	b 65.8		·	1.4	3.9	....	
Cowpeas (*Vigna catjang*) (Average of 13 analyses):c											
Original material....	13.0	21.4	....		1.4	b 56.7			4.1	3.4	....
Water-free substance	24.6	....		1.61	b 65.17			4.74	3.91	....	
Dried beans (average of 11 analyses):d											
Original material....	12.6	22.5	....		1.8	b 55.2			4.4	3.5	....
Water-free substance	25.74	....		2.07	b 63.16			5.03	4.0	....	
Dolichos sesquipedalis (green pods):											
Original material....	79.92	4.54	3.94	1.50	.53	2.74	1.75	3.25	2.56	1.17	3.53
Water-free substance	22.63	15.14	7.49	2.64	13.66	8.72	16.24	12.74	5.81	17.56	
String beans (average of 5 analyses):d											
Original material....	89.2	2.3	....		.3	b 5.5			1.9	.8	....
Water-free substance	21.50	....		2.77	b 50.93			17.59	7.41	....	

a Reported by Church, Food Grains of India. London: Chapman and Hall, 1886, p. 161.
b Carbohydrates by difference.
c U. S. Dept. Agr., Office of Experiment Stations Bul. 28 (rev. ed.), p. 37.
d Ibid., p. 65.

Harrison and Jenman[2] report the composition of two varieties of *D. lablab*, the crinkled-podded Bonavist bean and the flat-podded

[1] New York Cornell Sta. Bul. 67. p. 197.
[2] Rpt. Agr. Work Bot. Gard. British Guiana, 1891-92, p. 70.

Bonavist bean. Their results are not quoted, since the beans were evidently a different species from those purchased in San Francisco.

Both fresh and dry, the Dolichos analyzed in this laboratory compares favorably with the ordinary American beans, and were found on trial to be a satisfactory substitute for these, provided only the tough skin was removed from the dry seeds, which was readily effected. This species seems worthy of a general trial.

VARIOUS WATER PLANTS.

Nelumbium speciosum.—In reviewing the numerous uses of the sacred lotus (p. 17), it was stated that the seeds were an article of food among Asiatic nations, and it is not surprising to find them on sale in San Francisco. Their peculiar shape, the remarkably mature condition of their embryos, and the peculiar form of the embryo itself, all render the seed easy of recognition. Two forms were obtained, the one being somewhat larger and more irregular in shape than the other, but it was impossible to ascertain whether they were the product of plants which differed materially from each other. The smaller and more commonly used of the two forms is known as "pak lin" and is designated by the characters 白蓮; the larger is known as "seung lin" (上蓮). In all instances the thick outer coating of the seed on sale in San Francisco had been removed, leaving only the inner kernel. Under favorable conditions of temperature and moisture the seeds purchased germinated rapidly, and in the course of a few months from the time of starting produced thrifty plants. Thus far the plants from the two varieties of seeds do not differ materially from each other nor from the commonly cultivated form of lotus. As yet none have produced flowers. The germination of the seeds is somewhat interesting from a botanical point of view and is figured in Pl. II, fig. 4.

The seeds are eaten by the Chinese either raw, boiled, or roasted, being used as we use chestnuts. They are also said to be used in soup, though this seems to be a general term used to designate any mixture that has been boiled. The author was also assured that a form of "arrowroot," i. e., starch, is made from the seed.

The dark green germ is decidedly bitter and must be removed before the seed becomes palatable. Its bitterness has given rise to the Chinese saying, "bitter as the plumule of the lotus seed."

Analyses of the two forms are given in Table 14, p. 41. In the sample of large seed the germs, which form 1.3 per cent of the weight of the entire seed, were removed before analysis, but in the case of the smaller form the entire seed was analyzed.

An inspection of the two analyses shows that the seeds contain a high percentage of nutrients. The protein content is considerably above that of the commonly cultivated cereals, but does not approach that of the leguminous seeds. Nearly all of it is of an albuminoid nature. As might have been expected, starch is the most important nonnitrogenous constituent, though small amounts of other carbohydrates are also present. The starch found in both varieties differs in form from that found in the roots of the lotus.

Euryale ferox.—This is another aquatic, closely related to the lotus both in habit of growth and in botanical affinities. It is designated by the characters 芡實. Like the lotus its seeds form a part of the complex dietary of the Asiatic races.[1] These seeds were occasionally met with in the Chinese stores, but only in a broken and imperfect condition, apparently the result of a mechanical process intended to crack them open. Nevertheless the broken seeds showed such a remarkable resemblance to those of our native *Nuphar polysepalum*, which also furnishes seeds used by the Klamath Indians of Oregon[2] for food, that the author was at once led to suspect their relationship, and a dictionary in the Cantonese dialect, by John Eitel, associates the character used for them with the name given. The seeds are much smaller than those of the lotus, more farinaceous, and their embryos show only a moderate degree of development. The characters used to designate them are quite different from those used for the lotus seeds and the name commonly used is "tsz shat," which, however, does not appear in any of the works descriptive of Chinese botany accessible to the author. They are used principally as an addition to soup, in much the same manner as we use barley, but are also said to be used for the manufacture of starch.

The accompanying analysis (Table 14, p. 41) shows that they possess a food value about equivalent to that of the cereals, the principal constituent being starch, which is present in the form of exceedingly small grains, as shown in Pl. III, fig. 5. Sugars are probably not present, as the small amount reported in the analysis may represent traces of starch which repeated filtration failed to remove from the aqueous extract of the seeds. But little information is accessible regarding adaptability of the plant to the conditions found in the United States. It is already in cultivation in the North Atlantic States and other regions as an ornamental aquatic.

Trapa bispinosa.—The curiously-shaped seeds of this plant, which may be aptly compared with the head and horns of a cow, are familiar objects in the Chinese bazaars, where they are sold as curiosities. The regular Chinese merchant, however, is quite sure to have them in stock during the spring months, and they are sold by him as an article of diet. In the bazaars they are known as the "horn chestnut." The Chinese name is "ling ko" (菱角). This is also the name by which they are known in the Chinese classics, as the seeds are sometimes included among the five food grains of China. The seeds do not keep well, and those which find their way into the bazaars are often entirely decayed within.

The fresh seeds have a kernel which, in consistency and taste, resemble the chestnut. They will germinate readily if placed in a vessel of water

[1] Bretschneider, Jour. China Branch Roy. Asiatic Soc., 25 (1890), p. 218.
[2] Coville, Contrib. U. S. Nat. Herb., 5 (1897), No. 2, p. 96.

and produce the curious floating foliage characteristic of this genus of plants. (See Pl. II, fig. 1, and Pl. III, fig. 4.)

Authors are at variance as to the correct name of the plant producing these seeds. Forbes and Hemsley[1] regard it as one of the forms of the more widely known European *T. natans* inasmuch as they find many intermediate forms between the Chinese and the European species. By other writers the plant is variously designated as *T. bispinosa*, *T. bicornis*, *T. cochinchinensis*, and *T. incisa*. The plants to which these terms are applied are probably identical. As a means of distinguishing it from *T. natans*, from which the seeds analyzed by the author certainly differ, the name *T. bispinosa* has been adopted in this report.

The composition of the two sorts of Nelumbium seed, of *Euryale ferox*, and *Trapa bispinosa* are shown in the following table:

TABLE 14.—*Composition of seeds of various water plants.*

	Water.	Protein.	Albuminoids.	Amids (by difference).	Fat.	Starch.	Cane-sugar.	Reducing sugars.	Crude fiber.	Ash.	Undetermined.
Nelumbium speciosum (large form):	Per ct.	Per ct.	Per ct.	Per ct.	Per ct.	Per ct.	Per ct.	P. ct.	P. ct.	P. ct.	P. ct.
Original material ...	8.72	16.64	15.47	1.17	2.44	51.64	4.09	2.41	3.15	3.63	7.88
Water-free substance		18.23	16.95	1.28	2.67	56.57	4.48	2.64	3.45	3.32	8.63
Nelumbium speciosum (smaller form):											
Original material ...	9.40	17.73	17.64	.09	2.96	40.63	9.55	......	2.95	4.15	12.63
Water-free substance		19.57	19.46	.10	3.26	44.84	10.54	......	3.26	4.58	13.94
Euryale ferox:											
Original material ...	13.10	9.79	9.79	......	4.90	68.07	1.59	......	.83	.54	1.18
Water-free substance		11.26	11.26	......	5.64	78.33	1.83	......	.96	.62	1.36
Trapa bispinosa:											
Original material ...	10.59	10.88	10.42	.46	.65	69.39	......	3.95	1.41	2.57	9.56
Water-free substance		12.16	11.65	.54	.73	67.53	......	4.42	1.58	2.87	10.70

The analysis here reported of the seeds of *Trapa bispinosa* is not typical as regards water content, since the seeds used had been kept for a considerable time before they were analyzed and were unusually dry. The average weight of the seeds is about 4 grams and in those used the kernel constituted about 64.3 per cent of the total. Their food value is slightly less than that of the seeds of the Euryale, but, judged solely by their composition, they nevertheless contain a high percentage of nutritive material.

CHINESE MILLET.

Though this grain was not found on sale in San Francisco, it has a widely extended use in China and Japan. The sample analyzed in this laboratory was given to the author by Professor Fryer, who states that it is more especially used in the northern part of China. He considers it a very valuable article of diet and believes this plant should be introduced into cultivation in the United States. The species (presumably Panicum) to which the sample belongs can not be determined until

[1] Index Floræ Sinensis, 1887, p. 311.

plants have been grown from it. Several species of Panicum and closely related genera are in cultivation in China. Church reports the composition of a considerable number of millets used as food in India, and discusses their food value at some length.

The composition of the sample is shown in the following table:

TABLE 15.—*Composition of millet.*

	Water.	Protein.	Albuminoids.	Amids (by difference).	Fat.	Starch.	Cane sugar.	Reducing sugars.	Crude fiber.	Ash.	Undetermined
Panicum sp.:	Per ct.	Per ct.	Per ct.	Per ct.	Per ct.	Per ct.	Per ct.	P.ct.	P. ct.	P. ct.	P. ct.
Original material....	11.60	8.56	8.08	0.48	2.64	67.47	0.06	0.00	3.32	1.79	4.56
Water-free substance		9.68	9.14	.54	2.94	76.32	.07	.00	3.75	2.02	5.16
Panicum italicum: a											
Original material....	12.04	7.40	6.64	.76	3.87	45.73	b 28.48		1.37	1.11	....
Water-free substance		8.43	7.56	.87	4.40	51.99	b 32.38		1.54	1.26	....
Panicum miliaceum: c											
Original material....	12.0	12.6	....	....	3.6		d 69.4		1.0	1.4	....
Water-free substance		14.3	....	....	4.1		d 78.9		1.1	1.6	....

a Reported by Kellner, Landw. Vers. Stat., 30 (1884), p. 42.
b Determined by difference.
c Reported by Church, loc. cit., p. 43.
d Carbohydrates by difference.

For the sake of comparison, two analyses, one of a Japanese-grown sample, the other of a sample grown in India,[1] are included in the table above. All three analyses confirm the statement that in composition this grain ranks with the more commonly cultivated cereals. It is possible that it might find an important place among our cultivated grains.

FRUITS, NUTS, AND FLOWERS.

Nephelium litchi.—This tree furnishes the well known "li chee" or "lai chi" nuts (荔枝), which are so frequently used as tokens of good will by the Chinese laundryman or vegetable peddler about the time of their New Year. The edible portion of the li chee nut is really the fleshy aril which immediately surrounds the smooth brown seed and is in turn surrounded by the thin warty shell of the fruit itself. The nuts may be obtained in the dry form and also preserved with sugar in cans. In the sample analyzed the aril formed 46 per cent of the weight of the dry seed, but these figures can scarcely be considered typical as the fruits were kept for some months before being used. The composition of "li chee" nuts is shown in Table 16, p. 45. As might be expected from their sweet taste, sugar, especially reducing sugar, is their most important constituent. Other nutritive substances are present in small amounts.

The tree has been a favorite subject for greenhouse treatment in temperate climates for many years and is also reported to tolerate the

[1] Food Grains of India, pp. 31-59. London, Chapman & Hall, 1886.

climatic conditions found in southern California in a satisfactory manner.

Nephelium longan.—The fruit of this tree, designated by the characters 龍眼圓, differs from that of *N. litchi* in being somewhat smaller and smooth, otherwise it closely resembles the "li chee" nut. However, it is considered inferior to the li chee by the Chinese. The sample was kept for some time before analysis, and hence the figures for water content and those based on this determination are scarcely typical. The aril formed 40 per cent of the entire seed. Its composition is shown in Table 16, p. 45. In composition it does not differ greatly from the "li chee" save in the relative amounts of reducing and cane sugars present.

Zizyphus jujuba.—This is the Indian fig largely cultivated in tropical Asia, Africa, and Australia, and is designated by the characters 紅棗. The dried fruits are used by the Chinese in the same manner as the two mentioned above and are also well liked by Europeans. The European species (*Z. vulgaris*), is, however, better known and furnishes the important constituent of the jujube paste of the confectioners' shops. The composition of the fruit of *Z. jujuba* is shown in Table 16, p. 45.

Canarium album.—The so-called Chinese olive is the fruit of several species of Canarium, *C. album* and *C. pimela* being probably the two species whose fruit is chiefly used. The Chinese olive is a fleshy drupe 3 to 6 centimeters in length, which contains a hard, triangular, sharp-pointed seed. (See Pl. 11, fig. 6.) Within this are found one or more oily kernels. The fruits, known as "pak lam" (白欖), are found on sale in San Francisco, either green or salted and dried. They are also sold in the form of certain highly flavored preparations which are not generally acceptable to the American palate. The flesh of the fresh, yellowish-green fruit, like that of the true olive, is somewhat acrid and disagreeable, and requires special treatment before it can be made palatable. Smith[1] says that the fruits are often preserved in salt, and are also added to wine to moderate or counteract its effect.

The sample analyzed by the author (see Table 16, p. 45) represents the fresh pulp of the fruit. The most important constituent is fat, which forms nearly one-fourth of the total nutritive material.

The kernel of the seed of one variety of Chinese olive is also sold under the name of "lam yen" (欖仁). The peculiar form of these seeds is shown in Pl. 11, fig. 7. They are covered with a thick, reddish skin, but have the consistency and general characteristics of many other oily seeds. Microscopical examination showed the presence of well-formed aleurone grains, the crystalloids of which are remarkably well developed. The analysis of the seeds (reported in Table 16) shows that in composi-

[1] Contributions Toward the Materia Medica and Natural History of China, p. 50. Shanghai, 1871.

tion they closely resemble walnuts and similar nuts. The fat present consists of a yellow fluid oil, which absorbs 83.17 per cent of iodin. When fresh, the kernels are as palatable as walnuts and other common nuts. Little is known as to the adaptability of Chinese olive trees to other than tropical climates. It is questionable whether many portions of the United States would furnish the conditions favorable to their growth.

Ginkgo biloba (Salisburia adiantifolia).—A detailed account of this tree seems scarcely necessary, since it is well known—by nurserymen, at least—throughout the United States, though its merits as an ornamental tree are by no means fully appreciated. The trees grow readily and have fruited abundantly as far north as Washington, D. C. The white, thin-shelled nuts (Pl. II, fig. 2), known as "pak ko" or "gink ko" (白菓), are eaten by the Chinese, being first roasted or boiled, like chestnuts. They are somewhat acrid in the fresh state, and have a very disagreeable odor. When roasted or boiled, their flavor is peculiar. They are said by Hanbury,[1] Smith, and others to assist in digestion and to be eaten after meals for this purpose by the Chinese.

The pulp surrounding the seed of the fresh fruit has been examined by Bechamp,[2] who obtained from it the entire series of homologous fatty acids, from formic acid to caprylic acid. He also noted the occurrence of starch in the endosperm, though a complete analysis of the seed does not seem to have been made. The nuts found in San Francisco were evidently unusually dry. The analysis (Table 16, p. 45) shows that by far their most important constituent is starch. This is somewhat remarkable, as it is the only instance of which record has been found in which starch has been observed in any considerable quantity in the seed of a coniferous plant.

Some experiments were conducted by a student in this laboratory to determine whether the seed contained a digestive ferment. Samples of fish flesh were treated under suitable conditions with a decoction of the seeds, but no digestive action was observed.

Hemerocallis fulva.—Still another and very unusual vegetable substance largely used as a flavoring ingredient by the Chinese consists of the dried flowers of *H. fulva*, the day lily of American gardens. This substance is known as "kam cham t'soi," or the "gold-needle vegetable," and is designated by the characters 金針菜. It is reported by Bretschneider[3] to be in common use. As noted on page 20, Davy[4] refers to the use of the flowers of *Lilium bulbiferum* and *H. graminea* as food by the Chinese. The composition of flowers of *H. fulva* is shown in Table 16, p. 45.

[1] Notes on Chinese Materia Medica, Scientific Papers, 1876.
[2] Compt. Rend. Acad Sci. Paris, 58 (1864), p. 135; Ann. Chim. et Phys., 4. ser., 1 (1864), p. 288.
[3] Jour. China Branch Roy. Asiatic Soc., 15 (1880), p. 110.
[4] Erythea, 6 (1898), p. 25.

When judged by their composition, flowers of *H. fulva* are seen to possess a rather high food value. They are used, however, rather as a condiment than as an article of diet.

The composition of the fruit of *Nephelium litchi* ("li chee" nuts), *N. longan*, *Zizyphus jujuba*, *Canarium album* (Chinese olive), pulp and seeds; *Ginkgo biloba*, "gink ko" nuts, and the dried flowers of *Hemerocallis fulva* are shown in the following table:

TABLE 16.—*Composition of fruits, nuts, and flowers.*

	Water.	Protein.	Albuminoids.	Amides (by difference).	Fat.	Starch.	Cane sugar.	Reducing sugars.	Crude fibre.	Ash.	Undetermined.
	Per ct.	Per ct.	Per ct.	P. ct.	Per ct.	Per ct.	Per ct.	Per ct.	P. ct.	P. ct.	P. ct.
Nephelium litchi (aril):											
Original material....	14.94	2.91	....	....	1.44	....	4.47	66.58	....	2.21	7.45
Water-free substance	....	3.43	....	....	1.63	....	5.25	78.27	....	2.60	8.75
Nephelium longan (aril):											
Original material....	10.94	5.01	....	....	1.04	....	37.50	27.34	....	2.31	15.86
Water-free substance	....	5.63	....	....	1.17	....	42.11	30.70	....	2.59	17.80
Zizyphus jujuba (pulp):											
Original material....	13.44	2.93	....	....	....	....	13.06	42.19	....	1.73	26.65
Water-free substance	....	3.39	....	....	....	....	15.09	48.74	....	2.00	30.78
Canarium album (pulp):											
Original material....	73.22	.77	.61	.15	6.55	3.16	.53	1.95	4.15	1.50	8.17
Water-free substance	....	2.86	2.28	.57	24.46	11.79	1.99	7.26	15.48	5.61	30.53
Canarium sp. (seeds):											
Original material....	5.71	16.44	16.44	.00	58.57	.00	.92	.00	3.20	5.16	9.00
Water-free substance	....	17.44	17.44	.00	63.18	.00	.97	.00	3.39	5.47	9.54
Ginkgo biloba (seeds):											
Original material....	47.34	5.90	5.18	.72	.81	33.90	3.58	.00	.88	2.00	5.58
Water-free substance	....	11.21	9.84	1.37	1.53	64.34	6.79	.00	1.68	3.81	10.60
Hemerocallis fulva (dried flowers):											
Original material....	15.70	10.11	....	....	3.42	5.98	30.51	12.40	8.74	3.64	9.50
Water-free substance	....	11.99	....	....	4.06	7.09	36.19	14.71	10.37	4.32	11.27

FUNGI AND ALGÆ.

Russula sp.—Several species of fungi are largely used in the Chinese dietary, and when dried are standard articles of trade in San Francisco. The one most largely used is a white-spored species of Agaricini, probably of the genus Russula. It is designated by the characters 冬 瓜. It is easily distinguished from the American edible species by the tough, coriaceous character of both pileus and stipe, the latter being short and eccentrically attached. In color it is yellow, except the top of the pileus, which is brown or purple. At least two other species of this group are on sale in San Francisco, but these are delicacies rather than staple articles of diet, some of them retailing as high as $2 per pound.

The composition of the Russula is shown in Table 17, p. 47, and presents some interesting features. The figures for protein are far below those which have been assigned to edible fungi of this type, being only one-third to one-half the amounts reported for *Agaricus campestris*, species of Tuber,[1] etc. Of the protein, about four-fifths is in the form of albuminoids.

The exact nature of the large amount of carbohydrates present is unknown, and would form an interesting subject for investigation.

Peziza auricula.—Most of the notes of the earlier writers on Chinese botany contain references to the use of a peculiar mushroom which grows on decaying wood. This is a species of Peziza, probably *P. auricula*. It is designated as "mo eh" (Loureiro) or "mu ery" (木耳), by Henry, the English equivalent of which would be "wood ears," and is used in large quantities in San Francisco. The composition of the fungus is shown in Table 17, p. 47. The protein content of this fungus is remarkably low; sugars seem to be entirely absent, but some starch-like substance of an unknown nature forms more than one-third of the total weight. The fungus is decidedly tough and woody, and apparently possesses no especially valuable food characteristics.

Nostoc commune flagelliforme.—Of the algæ, one form which occurs in black tangled masses resembling horsehair is largely used by the Chinese. This was identified by Prof. W. A. Setchell as *Nostoc commune flagelliforme*. It is designated by the characters 髮菜. The use of this plant as an article of food does not seem to have been heretofore reported. Treated with boiling water, it forms a gelatinous mass, and is used as a thickening medium for various culinary preparations, one of the most esteemed being a combination of this vegetable with dried shrimps. The analysis shows a high percentage of protein, but only insignificant amounts of other food constituents.

Other forms of algæ which were identified by Professor Setchell are *Porphyra perforate*, and *P. nereocystis*, both of which are collected in large amounts at the Bay of Monterey, and two other species of the same genus, presumably *P. suborbiculata* and *P. tenera*, which are cultivated in Japan.

Species of Porphyra are used in considerable quantities for food by the natives in southeastern Alaska. The algæ are collected, pressed together and dried, and when used, are shaved into warm water and boiled to a thick porridge.[2] A sample of this algæ analyzed in the chemical laboratory of the Connecticut (Storrs) Station, had the composition reported in Table 17, p. 47.

The composition of the fungi and algæ analyzed, together with analyses of fungi quoted for purposes of comparison, is reported in the following table:

[1] König, Chemie der menschlichen Nahrungs- und Genussmittel, 3. ed., I, p. 747.
[2] This information is furnished by W. H. Evans, Ph. D., of this Department.

TABLE 17.—*Composition of fungi and algæ.*

	Water.	Protein.	Albuminoids.	Amids (by difference).	Fat.	Starch.	Cane sugar.	Reducing sugar.	Crude fiber.	Ash.	Undetermined.
	Per ct.	Per ct.	Per ct.	Per ct.	Per ct.	Per ct.	Per ct.	Per ct.	Per ct.	Per ct.	Per ct.
Russula sp.:											
Original material	8.40	15.42	12.87	2.55	2.56	21.41	.00	6.63	12.97	4.54	28.07
Water-free substance	......	16.83	14.05	2.78	2.79	23.37	.00	7.24	14.16	4.96	30.64
Peziza auricula:											
Original material ...	10.38	4.14	2.93	1.21	1.36	34.68	.00	.00	27.04	2.11	20.29
Water-free substance	......	4.62	3.27	1.35	1.52	38.69	.00	.00	30.17	2.35	22.64
Agaricus campestris californis: a											
Original material ...	92.84	3.40	......	......	.07				.54	.76	
Water-free substance	......	47.47	......	......	.96				7.55	10.67	
Morchella esculenta: b											
Original material ...	89.54	c 3.06	......	......	.50		d 1.60		.91	1.08	
Water-free substance	......	29.8	......	......	4.80		d 15.3		8.7	10.4	
Nostoc commune flagelliforme:											
Original material....	10.58	20.93	......	......	1.19		.00	.00	4.07	7.50	
Water-free substance	......	23.41	......	......	1.33				4.55	8.39	
Porphyra lacinata: e											
Original material....	21.85	25.70	......	......	.17		f 37.68			14.60	......
Water-free substance	......	32.88	......	......	.22		f 48.22			18.68	......

a Reported by Dahlen, Landw. Jahrb., 4 (1874), p. 639.
b Reported by Mendell, Amer. Jour. Physiol., 1 (1898), p. 225.
c Total nitrogen 0.49, albuminoid nitrogen 0.37, and nonalbuminoid nitrogen 0.12 per cent.
d Approximation.
e Unpublished analysis from Chem. Lab. of Connecticut (Storrs) Station.
f Determined by difference.

MISCELLANEOUS SUBSTANCES.

In addition to the above materials there are a number of others which are used to a greater or less extent. The following were examined. Although they did not seem important enough to demand an analysis they are worthy of at least a passing notice.

The roots of an aroid, which seems to be *Amorphophallus rivieri*, are frequently observed on the vegetable stands in the Chinese quarter of San Francisco. Plants grown from them at Berkeley have thus far failed to produce flowers, so that their absolute identification is as yet impossible. Henry,[1] however, refers to the use of the plant by the Chinese, but uses a character to designate it which differs from that used in San Francisco. The roots, known as "kai chau u" (鸡爪芋), are exceedingly acrid, and the directions given when the samples were purchased were to boil for several days before using. They are said to possess medicinal or at least restorative properties, the roots being used especially by persons who need a light diet. The roots sold in San Francisco, which are figured in Pl. III, fig. 6, are of interest on account of the peculiar compound starch grains which they contain. The roots of the closely related *Conophallus konjak*, which are largely used by the Japanese in much the same manner, have already been analyzed by Kellner,[2] who believed them to consist largely of starch. The later investiga-

[1] Notes on the Economic Botany of China, p. 35. Shanghai, 1893.
[2] Landw. Ver. Stat., 30 (1884), p. 42.

tions of Tsuji[1] have shown that the carbohydrate material consists quite largely of mannane and contains no starch.

The rhizomes of common ginger, *Zingiber officinale*, are largely used by the Chinese in California and may be obtained in a fresh condition throughout the year.

Among green vegetables the common purslane, *Portulaca oleracea*, is sometimes seen on the vegetable stands, but it is not extensively used. Purslane is used to some extent as a pot herb in American households. Its suitability for this purpose has been pointed out by Coville.[2]

According to an analysis reported by Huston,[3] purslane has the following percentage composition: Water, 86.56; protein, 1.81; fat, 0.50; nitrogen-free extract, 6.49; crude fiber, 2.12; ash, 2.23. It is noted that the ash content is unusually high.

The young shoots of a bamboo, probably a species of *Arundinaria*, are also sold, either pickled in brine or canned.

Of seeds, those of the common watermelon are very commonly eaten, usually raw. Two varieties of sesame seed, one black, the other white, are used in soup as barley is used in American households. A peculiar variety of the almond and another of the chestnut are also imported from China.

Among fruits, dried persimmons are often met with. The persimmon seems to be well adapted to this method of preservation. The seeds of *Areca catechu*, or betel palm, are often sold on the street corners in the Chinese quarters of San Francisco. A combination of slices of this nut with lime and the leaves of a species of piper, as used by the betel chewer, may also be obtained. However, they could not be identified.

The Chinese drug stores also furnish some curious objects of a vegetable nature. Among these were identified, chiefly through the report of Hanbury[4] on the Chinese materia medica, the fruit stalks of *Hovenia dulcis*, the seeds of *Torreya nucifera*, *Quisqualis indica*, *Momordica (Muricia) cochinchinensis*, *Terminalia chebula*, *Sterculia platanifolia*, *Ipomoea hederacea*, *Chalmongra* sp., *Melia* sp., and the fruits of a species of *Gardenia*.

---

[1] Imp. Univ. Col. Agr. [Tokyo] Bul., Vol. 2 (1891), p. 103.
[2] U. S. Dept. Agr., Yearbook 1895, p. 213.
[3] Indiana Sta. Rept. 1897, p. 16.
[4] Scientific Papers, 1876, pp. 228-251.